T0332079

Biology and Feminism

This book provides a unique introduction to the study of relationships between gender and biology, a core part of the feminist science research tradition which emerged nearly half a century ago. Lynn Hankinson Nelson presents an accessible and balanced discussion of research questions, background assumptions, methods, and hypotheses about biology and gender with which feminist scientists and science scholars critically and constructively engage. Writing from the perspective of contemporary philosophy of science, she examines the evidence for and ethical implications of biological hypotheses about gender, and discusses relevant philosophical issues including understandings of scientific objectivity, the nature of scientific reasoning, and relationships between biological research and the scientific and social contexts in which it is pursued. Clear and comprehensive, this volume addresses the engagements of feminist scientists and science scholars with a range of disciplines, including developmental and evolutionary biology, medicine, neurobiology, and primatology.

LYNN HANKINSON NELSON is Professor Emerita of Philosophy at the University of Washington. She has published *Who Knows* (1990) and numerous articles on feminist science scholarship. She is co-author of *On Quine* (1999), and is co-editor of *Feminism, Science, and the Philosophy of Science* (1997) and of a special issue of *Hypatia* on Feminist Science Studies (2004).

Cambridge Introductions to Philosophy and Biology

General editor
Michael Ruse, Florida State University

Associate editor
Denis Walsh, University of Toronto

Other titles in the series
Derek Turner, *Paleontology: A Philosophical Introduction*
R. Paul Thompson, *Agro-technology: A Philosophical Introduction*
Michael Ruse, *The Philosophy of Human Evolution*
Paul Griffiths and Karola Stotz, *Genetics and Philosophy: An Introduction*
Richard A. Richards, *Biological Classification: A Philosophical Introduction*

Biology and Feminism

A Philosophical Introduction

LYNN HANKINSON NELSON

University of Washington

CAMBRIDGE
UNIVERSITY PRESS

CAMBRIDGE
UNIVERSITY PRESS

University Printing House, Cambridge CB2 8BS, United Kingdom

One Liberty Plaza, 20th Floor, New York, NY 10006, USA

477 Williamstown Road, Port Melbourne, VIC 3207, Australia

4843/24, 2nd Floor, Ansari Road, Daryaganj, Delhi – 110002, India

79 Anson Road, #06–04/06, Singapore 079906

Cambridge University Press is part of the University of Cambridge.

It furthers the University's mission by disseminating knowledge in the pursuit of education, learning, and research at the highest international levels of excellence.

www.cambridge.org
Information on this title: www.cambridge.org/9781107090187
DOI: 10.1017/9781316106280

First published 2017

Printed in the United Kingdom by Clays, St Ives plc

A catalogue record for this publication is available from the British Library.

Library of Congress Cataloging-in-Publication Data
Names: Nelson, Lynn Hankinson, 1948- author.
Title: Biology and feminism : a philosophical introduction / Lynn Hankinson Nelson.
Other titles: Cambridge introductions to philosophy and biology.
Description: Cambridge, United Kingdom ; New York, NY : Cambridge University Press, 2017. |
Series: Cambridge introductions to philosophy and biology | Includes
bibliographical references and index.
Identifiers: LCCN 2017019180| ISBN 9781107090187 (hardback : alk. paper) |
ISBN 9781107462038 (pbk. : alk. paper)
Subjects: | MESH: Developmental Biology | Feminism | Biological Evolution | Philosophy
Classification: LCC QH491 | NLM QH 491 | DDC 571.8–dc23 LC record available at
https://lccn.loc.gov/2017019180

ISBN 978-1-107-09018-7 Hardback
ISBN 978-1-107-46203-8 Paperback

For my granddaughters, Charley Grace and Elyse Cate O'Connor

Contents

Figures

Every effort has been made to contact the relevant copyright holders for the images reproduced in this book. In the event of any error, the publisher will be pleased to make corrections in any reprints or future editions.

Preface

In the preface to the first edition of *The Selfish Gene*, Richard Dawkins identified three kinds of reader he had in mind as he wrote it. I also had groups of readers in mind and they helped shape the level, foci, and content of the forthcoming discussion.

The first and most important group of readers are senior undergraduate and graduate students who are assigned this text for a course, or who find it on their own, and are largely unfamiliar with feminist engagements with the biological sciences and/or without an extensive background in biology or philosophy of science. Accordingly, I have worked to present relevant topics and arguments with as much clarity as possible, and to include cases of feminist engagements with biology that are not only representative but also accessible to these readers.

The second and third groups of readers are those I felt, to paraphrase Dawkins, "looking over my shoulder" as I wrote: those with substantial background and expertise in some or even all the topics discussed here. Let's put feminist scientists and science scholars in the second group, and biologists and philosophers of biology in the third. But let's remember that many of those whose work we will study are both biologists and feminists, or both philosophers of biology and feminists.

I separate the groups because the "voices" I had in mind reflect different backgrounds and/or interests. Although I don't always follow the advice of those I think of as looking over my shoulder, at least not in the ways or to the extent they might have me do, keeping them in mind has led to this being a better discussion than it would have been had I not done so.

Acknowledgments

Writing a book can be a lonely process, but it is also one that, at least in my case, relies heavily on those whose work I have studied and learned from, and on the support of others. Colleagues in the Department of Philosophy at the University of Washington, particularly members of the Philosophy of Science and Feminist Philosophy reading groups, provided support and critical feedback, as did graduate and undergraduate students majoring in Biology and/or Philosophy, who read parts of the manuscript. My undergraduate research assistant, Sarah Weigelt, helped me to clarify issues so that students could understand them. Over many years, I have received support from feminist philosophers and scientists, and other philosophers and scientists, whose work has inspired me and whose constructive criticisms have been invaluable.

My husband, philosopher Jack Nelson, helped make this book come about in every way possible, reading and providing insightful comments about each chapter. My daughter Rebecca O'Connor's love, wise counsel, full support, and sense of humor have, as always, sustained me as I wrote the book.

A grant from the University of Washington's Royalty Research Fund supported my research and writing in 2014, for which I am very grateful.

Hilary Gaskin at Cambridge University Press consistently provided encouragement and support. Michael Ruse, general editor of the series of which this book is part, prodded and encouraged me when I felt discouraged (it is hard, I learned, to write a book that neither presupposes knowledge of the science discussed nor a familiarity with the philosophical issues raised by it, yet also does not oversimplify or misrepresent the science or philosophical issues). He provided constructive feedback during every stage of the project, although it will be clear we don't agree about everything.

I dedicate this book to my four-year-old granddaughters, with love and in the hope that by the time they are able to read it, many of the issues it discusses will have been resolved.

1   Conceptual Preliminaries

Introduction

This book is concerned with the question

> What, if any, are the connections between the biological sciences (evolutionary biology, neurobiology, primatology, developmental biology, and so forth) and feminism?

And with the question

> If there are any such connections, of what significance are they?

Before we can address these questions, we need an initial understanding of what feminism is, and of what kinds of connections between the biological sciences and feminism would be important to both. To be important, such connections would have to be substantive and empirically based. It is reasonably clear what areas of study fall within the biological sciences. Just what constitutes feminism is fairly complicated and so less obvious. It is a topic we will explore in some detail in subsequent chapters as we discuss specific connections alleged to obtain between feminism and biology. But there are some things we can say here about feminism that will be helpful in setting the stage for the discussions of this and later chapters.

Many readers may think of feminism as a purely political movement – a movement that seeks to identify and remove barriers to women being full participants in all aspects of life (social, political, economic, scientific, and so on). But feminism is more than a political movement. It is also a field of study – one subfied of which is devoted to gender and science that has, over the last 45 years, evolved into a loosely-delineated, multifaceted research program. Feminist scientists and science scholars claim to have shown that many disciplines, including many sciences, have relied on, and in many cases, continue to rely on, unquestioned, unwarranted, and, at times also unrecognized, assumptions about gender. Feminists have argued that the research programs in which such assumptions function are in need of reexamination. Feminist researchers have also developed alternatives to assumptions, research questions, and hypotheses that they criticize. Feminists engaged in this research include field and laboratory scientists, philosophers of science, historians of science, and scholars in a variety of other disciplines.

The general focus of feminist research devoted to science, a research program that has come to be called "Feminist Science Scholarship" or "Feminist Science Studies," is on the nature, sources, and roles of culturally and historically specific assumptions about gender in various sciences. We will be concentrating on feminists' arguments about the role and consequences of such assumptions in the biological sciences. It is quite easy to show that such assumptions have led to hypotheses and theories that maintain that there are important differences between women and men – including in cognitive abilities and temperament – and even to hypotheses that men are superior to women in terms of some abilities and characteristics.

To be sure, many theories claiming that men are superior to women in one way or another are no longer viewed as credible; but feminists argue that they are not just of historical interest. For one thing, feminists argue, they provide insights into general relationships between the biological sciences, on the one hand, and the social and scientific contexts within which biological research was undertaken, on the other hand. Studying research that is now recognized as flawed because it is or was based in part on unwarranted assumptions about gender can also help us see how such assumptions can shape, limit, or contribute to research questions, methods, and, ultimately, purported findings of present day biological research.

A less obvious, and more subtle and widespread, phenomenon that feminists study, and to which they seek to draw attention, is the influence of androcentrism, or male centeredness, in biological research that is not seeking to establish or explain that men are in some way or other superior to women. Research characterized by androcentrism, feminists argue, takes the activities, behaviors, dispositions, and the like typically (or at least stereotypically) associated with men or males as their primary focus, and fails to study (or adequately study) those typically associated with women or females. We will consider feminist arguments for the presence and consequences of androcentrism in a variety of biological sciences, as well as the constructive alternatives feminists within these sciences have proposed.

Another focus of feminist research is the role of "gender stereotypes" and "gendered metaphors" in the biological sciences. Readers are likely to be familiar with the general issue of stereotypes but may not be familiar with the second issue. Gendered metaphors attribute gender (together with characteristics commonly associated with the gender in question) to objects or processes that are not sexed. For example, traditional accounts of fertilization appearing in biology and medical texts portrayed the egg as "passive" and the sperm as "active." The egg's activism, in relation to the fusion of egg and sperm that yields a zygote, was not discussed (indeed, it is only relatively recently that it was even recognized). To cite another example, biologists often refer to specific hormones as "female" or as "male"; but of course, hormones are not organisms and such terms do not apply to them.

Feminists also focus on "equity issues" in biology, studying the formal barriers that once prevented women from entering it; and the potential consequences of relatively homogenous science communities in terms of gender, race, and class, on the questions asked, methods employed, and hypotheses proposed. Feminists also seek to identify remaining "informal" barriers to women's full participation and success in biology as well as other sciences.

Finally, feminists explore whether the lifting of formal barriers and lessening of informal barriers that led to increases in the number of women engaged in specific fields in biology has had an impact on the directions or content of the research and theorizing undertaken in them. Here it is important to note that most feminists do not view the differences of interest to involve the *sex* or *gender* of researchers, but rather in what ways feminist and feminist-friendly approaches reveal unwarranted assumptions about gender in scientific practice, and how, in the case of areas of biology, such approaches have made a difference. The earlier reference to the monolithic science communities common in the past is understood by feminists in terms of the monolithic experiences and perspectives that characterized them, not the gender, race, socioeconomic status, and the like, of individual scientists. Women trained in a science characterized to some degree by androcentrism might well accept the research priorities, categorizations, methods, and hypotheses of their field. We will also find that there are male scientists who approach their subjects of study in ways that are compatible with those of their feminist colleagues and that some describe themselves as feminists.

Readers will find that evolutionary theory is emphasized; three chapters focus directly on it (Chapters 2, 3, and 8). In addition, Chapter 4, which is devoted to primatology, discusses longstanding interests in that field to gain insights into "human nature" by studying nonhuman primates. Indeed, we will see in the discussion of evolutionary biology in Chapter 3, research devoted to identifying aspects of "human nature" is often regarded as an important aspect of evolutionary theorizing. In Chapter 6, which is devoted to medicine, we briefly consider Evolutionary Medicine. Finally, hypotheses about gender differences that we consider in the remaining chapters often trace some of their roots to evolutionary biology. Such relationships are not surprising. A common view (though it is not without its critics) was reflected in the title of a 1973 article by Theodosius Dobzhansky: "Nothing in biology makes sense except in the light of evolution" (Dobzhansky 1973).

Science Scholarship

Before proceeding, some comments about the nature of science scholarship (the study of science) are appropriate. Serious science scholarship, whether undertaken by philosophers of science, other science scholars, or scientists themselves, is committed to the view that proffered scientific claims are to be evaluated based on the *arguments and evidence* offered for them; and, for many such scholars (but not all), they are to be understood *in relation to the contexts* (scientific, *and* social, and historical) in which the claims are put forward. (Some contextualists only recognize contexts internal to science as consequential.) As many feminists whose analyses we consider do emphasize all such contexts, as well as relevant arguments and evidence, the forthcoming discussion will do so as well. "Contextualism," so understood, is initially explicated in Chapter 2 when we consider Darwin's hypothesis of sexual selection and his assumptions about sex and sex/gender differences. It is further discussed in later chapters, including different versions of it that feminists advocate. Feminist versions of Contextualism obviously do not accept the assumption that factors in the broader social context are always irrelevant to those how scientists evaluate hypotheses or to the content of science, including biology. And, as we will shortly see, Contextualism of any sort is controversial in those quarters in which an alternative approach, often called "Objectivism," is maintained. We consider Objectivism later in this chapter.

Philosophical Issues

Here and in subsequent chapters, we introduce philosophical issues that have been of interest in philosophy of science and are relevant to feminists' engagements with biology. In this section, we take note of some of these philosophical issues but defer detailed discussion of them to subsequent chapters. We adopt this approach because the philosophical issues here introduced, as well as others introduced in subsequent chapters, are abstract and complex, and best explicated and illustrated in relation to one or more areas of research. As appropriate, earlier chapters anticipate where an issue is discussed in more detail, and subsequent chapters refer back to where a detailed discussion of a relevant philosophical issue initially takes place.

Objectivity

Objectivity is so commonly associated with science that the fact that it has different meanings is often not noted outside the context of science scholarship. As philosopher Elizabeth A. Lloyd points out, objectivity can be and often is attributed to three quite different things even when its scope is limited to science. "Objective" is sometimes attributed to scientific hypotheses and theories that are taken to be true; sometimes used to describe the attitude to their research scientists are thought or expected to have, including detachment or lack of bias; and sometimes to describe "publicly available or accessible" facts (Lloyd 1993, 353). Often attributions of objectivity, and objectivity itself, are defined in terms of "value-freedom": that scientific theorizing and knowledge claims are not informed by values, and/or that scientists' research is not motivated by values. As we will see, many philosophers and scientists, including but not limited to many feminists, have come to question the possibility of "value-free science" and some also question whether values are inherently compromising to scientific inquiry. These views and other issues we will consider have led some feminists to attempt to reconceptualize "objectivity." We discuss the issues involved and such reconceptualizations in detail in Chapter 4. By then, we will have considered research questions and hypotheses in several biological sciences, as well as feminists' critiques of aspects of them and alternatives they propose.

The "Context of Discovery" versus the "Context of Justification"

Another philosophical issue relevant to our discussion involves a distinction philosophers introduced in the 1930s between what they called "the context of discovery" and "the context of justification." In brief, the context of discovery was taken to involve how hypotheses and theories come to be proposed; the context of justification was taken to be concerned with how hypotheses and theories are tested, and accepted, refined, or rejected. Proponents of the distinction have maintained that *how* a hypothesis comes to be proposed is not important, at least not to scientists or philosophers of science interested in understanding scientific reasoning and practice. What is important is how a hypothesis fares in the context of justification (e.g., Hempel 1966). From this perspective, many of the issues feminists raise about relationships between social beliefs and values, on the one hand, and research questions and hypotheses in biology, on the other, might easily be viewed as involving only the context of discovery and, as such, having no implications for understanding scientific reasoning and practice.

But feminists argue that the cases on which they have focused and we will consider demonstrate that the assertion that the context of discovery is unimportant or irrelevant to scientific reasoning is untenable. Problems such as androcentrism, they argue, often impact the *content* of science because they carry over into the context of justification. Historically and culturally specific assumptions about women and men, feminists argue, have had significant consequences for research priorities, research questions, hypotheses, observations, and the interpretation of test results – and that this is certainly the case in areas of biology. We consider many hypotheses and evaluations of them that feminists maintain demonstrate this. Obviously, the issues raised in arguments for and against the distinction are related in ways we will also consider to the issue of scientific objectivity.

Individualistic versus Social Accounts of Scientific Reasoning and Knowledge

For most of its history (at least from the time of Plato through the mid twentieth century), philosophy has focused on individuals in its analyses of knowledge. Philosophical analyses of science emphasized the role of logical reasoning – attributed to scientists *qua* individuals – and the role of the sensory experiences of individual scientists in scientific inquiry.

In the latter half of the twentieth century, the individualistic view of scientific reasoning began to be challenged. Philosophers and others, including social scientists, historians of science, and feminist philosophers, argued that scientific theorizing is inherently social. It is undertaken, they pointed out, within, and is informed by, the assumptions, research questions, and theories, of specific scientific communities. Feminist science scholars we noted also argue that social contexts external to science within which science is undertaken often have an impact on the directions and content of scientific research. Proponents of these several views have called for the development of accounts of the epistemology of science that study the social factors that are part of scientific reasoning and practice, and many have undertaken such studies. What has come to be known as "social epistemology" is not in fact limited to studying the social nature of scientific reasoning and knowledge, but of all knowledge. In forthcoming chapters, we discuss arguments, albeit arguments that sometimes differ in their details, for the view that scientific reasoning and knowledge are inherently social, paying special attention to those feminist scientists and science scholars offer. As we will see, most feminists who offer such arguments reject the idea that they entail relativism (the view that there are no grounds for evaluating hypotheses and observations) and view the social nature of science as calling for a reconsideration of traditional views of scientific objectivity.

Good Science versus Bad Science

For reasons anticipated in the philosophical issues so far discussed, feminists argue that the reasoning and hypotheses in biology they criticize and propose alternatives to are not plausibly written off as "bad science." And we will find in our discussion of specific hypotheses feminists critically engage, that they are or were in keeping with the methods, research questions, priorities, data, and hypotheses many accepted in the field in which they emerged. But we will also consider arguments offered by critics of feminist science scholarship that the hypotheses feminists engage are, in fact, bad or unsuccessful science and, thus, provide little if any insights into scientific reasoning or knowledge. Obviously, this is one of the more significant and complex issues raised in and by feminist engagements with biology.

Sex/Gender

One would hope (perhaps even expect) that the terms "sex" and "gender" so crucial to some areas of biological research and feminist science scholarship would enjoy unambiguous and generally accepted meanings. But this is not the case. In the various literatures that we consider, these terms are sometimes used interchangeably, sometimes taken to be clearly and importantly distinguishable, and sometimes understood as in need of further analysis. In the 1960s, psychologist John Money and others insisted on a sharp distinction between sex and gender as attributed to individuals. "Sex," they argued, should be used when describing the biological features of men and women. Gender, or "gender identity," results from social and cultural factors rather than biology. For obvious reasons, many engaged in the second wave of the Women's Movement embraced Money's distinction. Gender expectations, gender roles, and divisions in power along the lines of gender, they argued, are not a consequence of biology; rather, they are socially constructed and alterable.

We have so far focused on how feminist science scholarship explores relationships between *gender* and *science*, including biology. This is in part because many overviews of that scholarship, and many specific arguments offered by feminists, refer to issues involving "gender and science." But reexaminations of the distinction between sex and gender have been and continue to be part of feminist theorizing. The editors of a recent collection devoted to feminism and neuroscience summarized the results of such reexaminations, and concluded that a sharp distinction between sex and gender does not hold up when scrutinized (Bluhm et al. 2012). It is certainly the case that "sex differences" in behavior, abilities, temperament, and the like have long been attributed to biology, including alleged differences between men's and women's brains. But, the editors argue, because the body, including the brain, is changed by experiences and environmental factors (changes we will discuss in some detail in forthcoming chapters), the assumption that biology, including sex, is innate and stable, is unwarranted. And in relation to issues we will consider, some feminist biologists argue, as the editors put the point, that "it is impossible to disentangle the effects of sex from those of gender" on differences between women and men, including brain differences (ibid., 4).

Given these complexities, and because scientists and others whose work we consider often differ in how they use the terms "sex" and "gender," we will use "sex/gender" when discussing humans and "sex" when discussing other species, unless the research we are discussing explicitly defines one or both. Readers may initially find the term "sex/gender" awkward, but the research and arguments we will consider indicate that, often, it is appropriate.

There is also no consensus about the appropriateness or content of gender characteristics taken to be denoted by "masculinity" and "femininity," although we will see that these notions figure conspicuously in some biological research. Nor is there consensus or agreed-upon criteria about which entities (including but not limited to organisms) are appropriately described as "gendered." We will note differences in how the terms "masculinity" and "femininity" are used, and disagreements about their appropriateness, as they arise in the contexts of specific research programs.

Other Topics and Issues

As earlier noted, there is consensus among feminists that there are relationships between "equity issues" and issues involving androcentrism, gender stereotypes, gendered metaphors in the biological sciences, and other issues of concern to feminist scientists and science scholars. As we next briefly explore, and analyze in more depth in forthcoming chapters, the very idea that such relationships could or do obtain conflicts with longstanding beliefs about science's intersubjectivity and its autonomy from the beliefs and values characterizing the larger social contexts within which it is undertaken.

One such belief is that "who" is theorizing has no bearing on the content of science when things are going as they should. One argument taken to support this conclusion, but it is only one such argument of those we will consider, appeals to the purported distinction between the contexts of discovery and justification. So, for example, the fact that for many generations most scientists were white males, and that the larger social context within which they worked was characterized by sexist and/or androcentric beliefs, values, and policies, might have had an impact on the hypotheses they proposed – but *not* on the most important issue: how those hypotheses were tested and ultimately judged. From this perspective, scientific methods, including experimental tests, and scientific norms, can and will insure that androcentrism and/or sexism do not affect the content of science.

A second such belief can be aptly described as encompassing two norms: first, that science must be allowed to pursue knowledge without interference based on religious, political, or other "nonscientific" views or preferences; and second, that science can and should (and some would add "does") keep itself "aloof from" non-epistemic values – that is, from any values that are unrelated to the pursuit of knowledge (e.g., Quine 1981, 49). Episodes such as The Copernican and Darwinian Revolutions underscore the rationale for the first norm. And the kinds of scientific theorizing about racial differences that occurred during the nineteenth and early twentieth centuries serve as an example for the rationale of the second norm.

Do feminists' engagements with biology violate one or both norms? As will become obvious in forthcoming discussions, most feminist scientists and science scholars – and certainly those whose work we consider – do not view their research as motivated by a desire to *undermine* science. To the contrary, their analyses of biological research are generally characterized by frequent and substantive appeals to science, including empirical evidence. They use scientific methods, data, and empirical hypotheses to challenge the assumptions and/or hypotheses they criticize. Nor do most feminists believe that the scientists whose work they criticize are guilty of conscious bias, overtly manipulating data, or misrepresenting experimental results. Research in which such things occur is patently uninteresting precisely because it provides few if any insights into the actual nature of the biological sciences or their relationship to social beliefs and values. As philosopher and archaeologist Alison Wylie makes this point, to engage in critiques of the sort feminists undertake, is to "restudy" some aspect of scientific research (Wylie 1997).

In addition, as earlier noted and as we will subsequently explore in detail, for the most part feminists' engagements with the biological sciences are by no means limited to critiques. Their engagements are frequently constructive – offering alternatives to the research and hypotheses on which they focus. Moreover, we will study specific fields in which the alternatives they recommend to traditional approaches have been accepted and have come to characterize the field in question. The qualification – "most feminist scientists" – within the previous paragraph reflects the fact that there are exceptions to these generalizations. Most of these are to be found in the early days of feminist attention to biology and gender, and more recently in postmodernist critiques of science. In the research on which we focus, such approaches are rare.

We now have at least a rough and preliminary account of the nature of feminist research concerning science, including biology. But some readers might object that the kind of feminist research so far described is uncalled for and unnecessary. These readers may hold that science is quite different from other academic disciplines (for example, literary criticism, political theory, and perhaps most obviously, philosophy itself). They may well find the following characterization of science, often referred to as "Objectivism," more in keeping with their own views or experiences of science and more warranted.

Objectivism

According to Objectivism, science is

- concerned with the facts (and only the facts) that, once discovered, are in some sense, obvious and incontestable;
- undertaken by investigators who, using "the scientific method or methods," gather these facts, analyze them, draw only warranted inferences from them, and bring no presuppositions or background assumptions to these activities; and therefore
- is characterized by a degree of "universality" in the sense that "who" is doing the science and in what context it is done do not affect what research is undertaken or the outcomes of that research.

Given this characterization of science, it might well seem inconceivable, or at least highly unlikely, that there could be any need for feminist engagements with the biological sciences, or indeed any connection whatsoever between any science and feminism.

And it is true that science has often been, and sometimes still is, portrayed as above – in K–12 and university textbooks and classrooms – and sometimes by scientists when they respond to what they view as misguided and non-scientific attacks on science. An example of the latter is Nobel Laureate physicist Sheldon Glashow's response to critical analyses of science, including some leveled by feminists, at a 1987 conference on the future of science:

> We scientists believe that the world is knowable, that there are simple rules governing the behavior of matter and the evolution of the universe. We affirm that there are objective, extra-historical, socially neutral, external and universal truths, and that the assemblage of these truths is what we call

physical science. Natural laws can be discovered that are universal, invariable, inviolable, genderless, and verifiable.

They may be found by men or by women, or by mixed collaborations of any obscene proportions. Any intelligent alien anywhere would have come upon the same logical system as we have to explain the structure of protons and the nature of supernovae.

This statement I cannot prove. This statement I cannot justify. This is my faith. (Glashow 1987)

Glashow clearly does hold a view of science like the one described above as "Objectivist." But it might seem odd that he takes his commitment to science so conceived as a matter of faith rather than reason and evidence. For if his view is a matter of faith, then others may be equally justified in holding alternative views as a matter of faith – for example, that all the important truths about nature are found not by science, but by divine revelation. A more plausible explanation is that Glashow merely intends to acknowledge the fallibility of science – to acknowledge that while science has been enormously successful, any given claim of science could turn out to be mistaken.

And this, of course, is well recognized by scientists. Focusing on his own field, the physicist Pierre Duhem argued early in the twentieth century that physicists can never be certain that there is not an alternative hypothesis or theory as yet undiscovered that might be recognized in the future as more warranted than one currently maintained (Duhem 1914). In addition, as the philosopher W.V. Quine argued beginning in the 1960s, nor could we or will we collect *all* of the evidence needed to *prove* any empirical hypothesis or theory. There is, as he put the point, "empirical slack" between all the sensory evidence available to us and the generalizations that make up hypotheses and theories in common-sense theorizing as well as in science.

The truths that can be said even in common-sense about ordinary things are themselves . . . far in excess of any available data. [There is a] basic indeterminacy . . . events are less determined by our surface irritations. This remains true even if we include all past, present, and future irritations of all the far-flung surfaces of mankind. (Quine 1960, 22)

This view is one line of argument for what has come to be called "the underdetermination thesis," a thesis we will have more to say about in later chapters. A second line of argument associated with the thesis and offered by

Duhem and Quine (although many view their arguments as different in important ways) is that individual hypotheses do not, in Quine's words, "face the tribunal of experience" in isolation – but only as part of broader bodies of theory. We will not focus on this second line of argument, although there are feminists who have done so in their analyses of scientific reasoning (e.g., Nelson 1990).

Although Objectivism has its advocates, it is no longer accepted by many philosophers and historians of science, or by all scientists. Moreover, even a cursory study of the history of science challenges core aspects of the view. As examples, we briefly consider three cases from the history of science. None is an instance of "bad" science, but all do involve a reliance on unexamined and/or untested assumptions (call these "background assumptions"); in addition, all illustrate the contextualization of science: that is, they show how the historical contexts, in some cases internal to science and in others, external to science, within which science is done, can influence scientific reasoning.

The Role of Background Assumptions in Scientific Reasoning: Three Examples

Tycho Brahe and Stellar Parallax

Our first example, research undertaken by the astronomer Tycho Brahe (1546–1601) that led him to reject the Copernican hypothesis that Earth revolves around the sun, is not related to either feminism or biology. Brahe was an eminent astronomer and the first to make careful, accurate, and regular (nightly for over 20 years) observations of the stars and other heavenly bodies. Although he did not have a telescope (he used only a compass, a sextant, and various other instruments that did not involve lenses or magnification, many of his own design), his observations were far more accurate than any previously made.

The physics and astronomy Brahe learned were, respectively, those of Aristotle and Ptolemy. Ptolemy held that Earth is at the center of the universe and that the moon, sun, planets, and stars revolve around it. These heavenly bodies, Ptolemy held, are attached to eight globes or shells, and these shells revolve around Earth. The stars are all attached to the outermost shell, the other heavenly bodies to the seven inner shells. All these shells

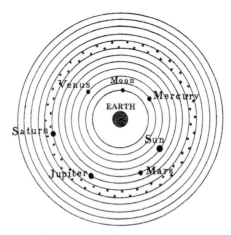

Figure 1.1 Ptolemaic Cosmology: Geocentric model of the universe.
Sheila Terry, Getty Images.

were taken to be made of a crystalline substance through which light could pass. Change, birth, death, growth, and decay, occurred only in the sublunar sphere, that is, below the sphere that carried the moon. The heavens, which made up the superlunar sphere, were taken to be eternal and unchanging (see Figure 1.1).

In 1543, the mathematician and astronomer Nicolaus Copernicus published *On the Revolutions of the Heavenly Bodies* (*De Revolutionibus Orbium Coelestium*), which argued that predictions of the movements of the planets were more accurate if one took the sun to be stationary and Earth to be a planet that revolved around it. Brahe knew there were problems with the geocentric model of the universe. Planetary positions and apparent movements were difficult to calculate. And Brahe himself had observed a supernova and a comet. Both were clear evidence of change in the heavens, although whether they took place in the sublunar sphere or beyond it was unclear. But they and the problems with predicting planetary movements suggested that, perhaps, the geocentric model of the universe is not wholly accurate. Brahe became interested in the Copernican model and used his extensive and meticulous observations to test it. The philosopher of science Carl Hempel describes Brahe's reasoning as follows:

> [I]f the Copernican hypothesis were true, then the direction in which a fixed star would be seen by an observer on the earth at a fixed time of day should gradually change; for in the course of the annual travel of the earth about the

sun, the star would be observed from a steadily changing vantage point – just as a child on a merry-go-round observes the face of an onlooker from a changing vantage point and therefore sees it in a constantly changing direction. More specifically, the direction from the observer to the star should vary periodically between two extremes, corresponding to opposite vantage points on the earth's orbit about the sun. The angle subtended by these points is called the annual parallax of the star; the farther the star is from the earth, the smaller will be its parallax. Brahe, who made his observations before the telescope was introduced, searched with his most precise instruments for evidence of such "parallactic motions" of fixed stars – and found none. (Hempel, 1966. 23–25)

In fact, stellar parallax does occur but was not observed until the nineteenth century when powerful telescopes came into use. Brahe was unable to observe it because the stars are enormously further from Earth than had been assumed. Brahe considered that this might be so, and hence that stellar parallax might occur; but he rejected the assumption that the stars were enormously further away and rejected the Copernican hypothesis that Earth revolves around the sun.

Why? One possible answer is that, because the Copernican hypothesis did not place Earth at the center of the universe, it was incompatible with Aristotle's physics – the only physics then available. Accepting the Copernican hypothesis would require a new physics, and Brahe didn't have a new physics. Another answer, not incompatible with the first, is that the ancients and all subsequent astronomers had assumed the stars were not a great deal farther from Earth than are the planets – perhaps a natural assumption given the crystalline shell model of the universe. Given this, adding the hypothesis that the stars are much further away might have seemed to Brahe to be an *ad hoc* hypothesis – as hypotheses are called when there is no evidence to support them and they are proposed only to save a theory or hypothesis – in this case, the Copernican hypothesis that Earth revolves around the sun. In the end, Brahe could not shake himself of the background assumption about the stars' distance from Earth. As a result, although he was a brilliant astronomer with highly sophisticated tools, he concluded incorrectly that stellar parallax does not occur. The important points here are that although Tycho Brahe recognized that he was bringing a background assumption that might be problematic to bear, he continued to use it; and that although he reached the wrong conclusion, Brahe's test of the Copernican hypothesis was *not* bad science. In fact, it was very good science.

Hempel cites the role of Brahe's background assumption – that the stars are not enormously distant from Earth, or as Hempel called it, Brahe's "auxiliary hypothesis" – in laying out his own thesis that background assumptions are always involved, whether they are recognized or not, in the evaluation of the evidence for a hypothesis. As Hempel points out, at the very least the test of any hypothesis assumes "ceteris paribus," which roughly translates to "all things being equal." That is, at a minimum, scientists must assume that there are no factors unknown to them that might affect the results of their tests (Hempel 1966). Although Brahe recognized that his assumption that the stars are not a great distance from Earth was just that, an assumption, he did not abandon it and did take his failure to observe stellar parallax to refute the Copernican hypothesis that Earth revolves around the sun. This is an illustration of how background assumptions can, and in this and other cases did and do, impact the design of the test of a hypothesis and the interpretation of the test's result.

Aristotle on Sex/Gender Differences and on Reproduction

Aristotle (384–322 BCE) was a logician, philosopher, physicist, and biologist, and without question, he excelled in each (as well as in other endeavors). His physics reigned until Galileo challenged it and alternatives to it were developed in the seventeenth century, and his logic is still taught. But like many scientists, his views were in part informed by his historical and social context. For Aristotle, this context was ancient Greece, and fundamental differences between men and women, and males and females generally, were widely accepted and argued for. In brief, from at least Plato (428–348 BCE) forward, it was believed that men are superior to women in myriad ways. Like other philosophers of his time, Aristotle theorized about the differences between the sex/genders and sexes, largely in his biological research. He also theorized about sexual reproduction, and the different contributions males and females make to it.

But he faced a conundrum putting together his theories concerning the "nature" of each sex/gender and his explanation of human reproduction. Women, Aristotle held, are "incomplete [or distorted] men," physically and intellectually inferior to them. And his explanation of sexual reproduction was androcentric (i.e., "male centered"), apparently reflecting the view that women's role, in all areas of human endeavor, was of far less consequence than that of men. Aristotle maintained that women provide the space and necessary physical matter for a human to grow. Sperm contain the "form" of a tiny human male. Given this one would expect all babies to be male. But this is obviously not the case. Aristotle hypothesized that, while the form a man contributes is always male, female offspring result when a woman's womb is insufficiently warm – resulting in an incomplete or distorted child – that is, a female.

Now this might strike readers as an *ad hoc* hypothesis, a notion introduced in our discussion of Brahe. But Aristotle did not consider his hypothesis about insufficiently warm wombs to be *ad hoc*. Neither he nor his contemporaries, as far as we know, had any understanding of female reproductive processes, including the existence of ovaries and eggs. In addition, gestation was not recognized as a period of *developmental change*, but simply a process of *growth*. Finally, Aristotle's appeal to insufficient warmth is consistent with his explanation of other aspects of women's biology. For example, he maintained that menstrual blood is female-produced semen and that it is not whitish, as is male semen, because it is insufficiently heated.

It appears that Aristotle's biological theories about the nature of males and females, including men and women, were in part reflective of the then culturally accepted but unproven "background assumption" that women are inferior to men. Aristotle did not examine or challenge this assumption, but rather allowed it to shape his biological studies of the sex/genders and sexual reproduction. It is also, we will see in subsequent chapters, a common background assumption (or, in some cases, a hypothesis) in later scientific periods, although not nearly as prevalent today except in so far as some scientists continue to maintain that men are superior in math and science, and offer biological explanations for their superiority, as we consider in Chapter 7.

Ignaz Semmelweis and the Cause of Childbed Fever

To illustrate and argue for the role of background assumptions in scientific reasoning, Hempel also used the reasoning underlying an experiment devised in the nineteenth century by the physician Ignaz Semmelweis to test his hypothesis about the cause of what was then called "childbed fever," a condition that was causing the death of many European women during or following childbirth, and often the death of their newborns (Hempel 1966, 4–6).

Semmelweis was fortunate enough to be able to reason within the context of "a natural experiment" – what we call an experiment that nature presents and that, in some cases (including that which Semmelweis was dealing with), no scientist or physician would engineer for ethical reasons. Semmelweis was the head of a medical facility that included two maternity wards and the death rate of childbed fever in them was significantly different.

After exploring differences between the two wards and concluding that most were not related to childbed fever, an accident involving a colleague led Semmelweis to hypothesize that exposure to cadaveric material caused the fever. His colleague, like Semmelweis and medical students at the facility, regularly performed autopsies and he nicked himself with the scalpel he had been using to do so. He developed the symptoms of childbed fever and died. Noting that in the ward with the higher rate of childbed fever, women were examined by medical students after the students had performed autopsies but either had not washed their hands or only casually washed them, and that those in the ward with the lower rate were examined only by midwives who did not perform autopsies, Semmelweis hypothesized that what he called "cadaveric material" was the cause of childbed fever.

As Hempel relates, to test his hypothesis, Semmelweis instructed medical students to wash their hands in a solution of chlorinated lime after performing autopsies and before examining women in their ward. To paraphrase Hempel, Semmelweis' reasoning took the following form:

1. If cadaveric material causes childbed fever, then if medical students wash their hands in a solution of chlorinated lime, the death rate in the ward in which they examine women will decrease.
2. Medical students will wash their hands in a solution of chlorinated lime.

==

3. The rate of childbed fever in the ward in which medical students examine women will decrease (ibid., 5–6).

Although subsequent events would indicate that cadaveric material was not the only cause of childbed fever, Semmelweis' initial test confirmed his hypothesis about cadaveric material: the rate of childbed fever did decrease when medical students washed their hands in the chlorinated lime solution before examining women in their ward.

To illustrate the role of background assumptions in Semmelweis' reasoning, Hempel asks us to consider the following thought experiment.

What if, despite the evidence provided by the death of his colleague and the apparent refutation (based on experiments or logical reasoning) of alternative explanations, the death rate had not decreased after the initial experiment (ibid., 17–20)? Would the logic or results of Semmelweis' experiment "force" or entail the conclusion that cadaveric material is not a or the cause of childbed fever? No, Hempel argues, given the role of the background assumption "chlorinated lime solution will kill whatever it is in cadaveric material that causes childbed fever." If the predicted outcome had not occurred, it could be that the hypothesis that cadaveric material causes childbed fever was wrong; but it could equally well be a problem with this background assumption. Perhaps *it* was wrong. Similarly, the experiment assumed that medical students *did* wash their hands in the solution. Perhaps they did not – a genuine possibility given that hand washing was not, at the time, recognized as important and that, initially, Semmelweis' hypothesis was rejected by many physicians.

Hempel's case study further illustrates the role of background assumptions, or as he called them, "auxiliary hypotheses," in scientific reasoning. As we noted, he argued in more general terms that *every* test of a hypothesis at least presupposes "ceteris paribus." As his examples involving Brahe and Semmelweis are intended to show, scientific reasoning also includes background assumptions of specific relevance to the case at hand.

We now have three examples of a good scientist doing good science *and* relying on unquestioned (or, in the case of Brahe, even though questioned, nonetheless maintained) and untested "background assumptions." As we will see, feminist biologists and philosophers often point to the role of background assumptions in theorizing in biology, including unwarranted background assumptions about sex/gender. Feminists frequently appeal to philosophers' arguments (e.g., van Fraassen 1980) that the "distance" between empirical evidence and a hypothesis is necessarily mediated by such assumptions.

What Should We Take from These and Similar Examples?

One might think it reasonable to respond to the cases from the history of science we have considered, and others like them that we will consider in forthcoming chapters, as follows. These examples do show that in the past science has sometimes made missteps. But that was then and this is now. Pre-contemporary science and/or pre-mature science (science before the advent of Ph.D. programs, scientific journals, professional associations, and so forth) might sometimes have appealed to unquestioned and untested assumptions, including assumptions about sex/gender. And these assumptions may indeed have been value-laden and the products of specific historical, cultural, and political contexts. So, too, before the emergence of discipline-specific norms for scientific research, of peer-review, of the recognition of the importance of replicability of results, and the like, the influence of unwarranted assumptions and/or current social beliefs and values on the directions and/or content of scientific research might well have been substantial. But surely in present twenty-first-century science (and much of twentieth-century science) cases like these are rare, and the influences previously at work are no longer a factor, or at most a minor or rare factor, in cases involving "good" science.

But "then" and "now" are moving targets. What once was current science is now past science, and what is now current science will *become* past science. Science is always evolving and there is no reason to think *this* point in time is different from *past* points in time, or from *future* points. Put another way, only substantive and empirical study (of a rather vast sort – and, even if possible, not yet undertaken) could justify the conclusion that unwarranted and unquestioned assumptions are no longer at work in science. Consider this: except as a case study in the history of science, no present day university professor would assign the science texts my generation were assigned as undergraduates and graduate students – before we had theories that included dark matter and dark energy, images of and lessons learned from the Hubble Ultra-Deep Field, evidence of a universe that is expanding much more rapidly than was previously thought, the theory of plate-tectonics, knowledge of numerous hominids that preceded or co-existed with *Homo sapiens*, genome projects, and so forth. The texts I studied are outdated, their accounts of the various sciences are in many ways not just incomplete by today's standards but also just wrong in various ways. It is not just that we now know more than we did then; but, also, that some of what we now claim to know conflicts with what we claimed to know then. But it does not follow that what was accepted and studied in the 1960s can and should now be recognized as "bad" or "unsuccessful" science. It was good science and it was successful, when measured by all the standard cognitive norms for good scientific practice. But, to repeat, science evolves and that it does is one of its signature features.

This feature of science is in fact reason to believe that not all the science we now accept will continue to be accepted in the future. That social beliefs and values have been infused in science in past periods suggests that they may be infused in today's science. The history of science is in part a history of discovering the beliefs and values internal and external to science that came to inform science. If it happened before it can be happening now and can happen in the future. The questions that need to be confronted and explored by scientists, science scholars, and by anyone who wants to understand science are: "In what ways, and to what degree, and with what consequences, do the sciences (in the case at hand, the biological sciences) reflect their historically and culturally specific contexts, both within science communities and within the broader social context?" and "What are the implications of answers to these questions for the nature of science and of scientific reasoning?"

Questions Feminists Ask about Biology

One way to provide content to the relationships between biology and feminism as it is currently studied and pursued is to consider a representative sample of questions feminist biologists and science scholars ask. We will explore them in the chapters ahead.

- For what period(s) and for what reason(s) were women unable to participate in biology? Are there any "informal barriers" to their full participation and success in one or more fields of biology that remain?
- Did an influx of women and/or feminists into a biological science in which they were previously underrepresented have any impact on research questions, methods, or hypotheses about sex/gender?
- What questions do biologists ask about the sex/genders and sexes, and about biological, behavioral, intellectual, and/or other differences between them? In what ways are their questions framed?
- What conclusions about the sex/genders, including differences between them, do or have biologists drawn based on their research? What is the evidence for these conclusions?
- What issues of concern to women are addressed in the biological sciences, and what issues are not?

- How, if at all, might a biologist's views about sex/gender influence aspects of her or his theorizing about issues that, at least on the surface, are unrelated to gender?
- How have, and how do, biological theories about women and men, females and males, reflected or reflect beliefs and values in the larger social context within which biological science is undertaken?
- How have, and how do, biological theories about women and men, females and males, reinforce or justify beliefs and values in the larger social context within which the relevant science is undertaken?

2 Sexual Selection

Darwin

Introduction

Evolution requires heritable variation (variation that is passed down from one generation to another), so it is not surprising that evolutionary theorists frequently study reproduction. They ask questions such as, "Why do so many species engage in sexual, rather than asexual, reproduction?" And they ask, "When did sexual reproduction first emerge?" The first question arises because organisms that reproduce asexually pass on all their genes to their offspring (making asexual reproduction appear to be the preferable method of ensuring one's genes are passed on), while the offspring of organisms that reproduce sexually inherit half their genes from each parent. In this and two later chapters, we discuss evolutionary theorizing about mating strategies (generally assumed by the research on which we focus to be different for females and males) and about other sex/gender differences that some evolutionary theorists have argued resulted from evolution and are significant – beyond those obviously related to sexual reproduction (i.e., that females produce eggs and males produce sperm, and anatomical differences related to reproduction).

Here we begin with the theorizing about such differences between the sex/genders and sexes in which Charles Darwin engaged in *On the Origin of Species* (1859) and *The Descent of Man* (1871). In Chapter 3, we continue with evolutionary theorizing about differences between the sex/genders and sexes that was undertaken in the twentieth century by advocates of Parental Investment Theory and Human Sociobiology between the 1960s and 1980s. In Chapter 8, we consider Human Evolutionary Psychology, a current research program. What links these theories is a commitment to sexual selection as an evolutionary mechanism and to sex/gender and/or sex differences in behavior, psychology, and mating strategies that are viewed to be related to or a consequence of it.

Philosophical Issues

As in Chapter 1 and later chapters, we devote sections of this chapter to specific philosophical issues, including some approaches to studying science as they arose in twentieth-century philosophy of science to which feminists appeal, as well as specific issues raised by feminists that are relevant to the science we are considering. We discuss the developments that led to the emergence of "Contextualism" as a general epistemological approach to studying science. We also discuss the thesis that observations, in science and so-called common sense theorizing, are "theory laden": that they are partly shaped by accepted theories, background assumptions, concepts, and/ or language. Neither issue originated in feminist engagements with biology or with science more generally, but we will see that feminists' analyses of science, including of Darwin's hypotheses, appeal to them.

Feminists' analyses of areas of research in biology also raise specific philosophical issues. In this chapter, we consider the role of "gender stereotypes" in scientific theorizing, focusing in particular on feminists' arguments that such stereotypes substantively informed Darwin's observations and theories. To call a descriptive word or phrase a "stereotype" is to challenge its appropriateness on the grounds that it conveys an inaccurate or unwarranted characterization of an entire group, the members of which may have far less in common than the stereotype suggests. Feminists claim, for example, that when Darwin (and other evolutionary theorists) describe males as "more eager to engage in sex" than females, and as "providers and protectors of females and offspring," they are assuming gender stereotypes. In later chapters, we consider feminist critiques of "gender stereotypes" in other areas of biological research.

In Chapter 3, we devote sections to feminists' critiques of biological or genetic determinism, a methodological and ontological assumption that often influences biological research that focuses on what are alleged to be sex/gender differences in psychology and behavior. We anticipate this issue here because although we will not engage in a sustained discussion of it until the next chapter, we will see that biological determinism informs Darwin's views about sex and sex/gender differences that we do discuss.

Charles Darwin on Sexual Selection, and Sex and Sex/Gender Differences

Charles Darwin (1809–1882) is generally credited with being the first (although not the only) scientist to identify "natural selection" as an evolutionary mechanism. (Alfred Russel Wallace (1823–1913) independently discovered it some twenty years later, just a year before Darwin published *On the Origin of Species* in 1859.) This "co-discovery," historians of science point out, is not as surprising as it might seem; the two men were reading the same books and considering what was by then widely disseminated information about embryological and anatomical similarities across quite different species, the geographical distribution of species, extinctions, and other phenomena taken to be evidence of evolution.

Darwin was also the first to propose "sexual selection" as a secondary evolutionary mechanism. The story about why Darwin became interested in differences between males and females is well known among Darwin scholars and, among them, understandable. For Darwin, the issue did not arise in a *scientific* vacuum (i.e., it was not solely related to the preoccupation of Victorian England with sex/gender differences), as will shortly become clear. Nor is there evidence to indicate that his attention to sex/gender differences was primarily motivated by a desire to establish that men are superior to women in terms of highly valued qualities, such as intelligence and leadership.

Rather, Darwin's interest in sex and sex/gender differences arose in response to what appeared to be a serious problem for his theory of evolution by natural selection – indeed, perhaps even a group of out-and-out counterexamples to it. At the same time, we will see that in his account of sexual selection, Darwin did state (and in some places, argue) that men are superior to women (and did maintain that, in general, males are superior to females) in terms of what were then and are still viewed as important characteristics. It is also true that his accounts of differences between men and women were well in keeping with widely held beliefs in the Victorian social and political context within which he was working. But, again, establishing male superiority and sex differences more generally was not what motivated Darwin's interest in sex and sex/gender differences.

To understand what led Darwin to pay attention to sex and sex/gender differences, we need to review his arguments in the *Origin* that natural selection is a mechanism that brings about evolution (all six editions of the work include statements that it is an important mechanism, but not the only such mechanism). His general argument for natural selection is paraphrased below and, as he noted, its two premises contained "no new facts." Each was generally accepted within the loosely delineated scientific community of his time (which was before the institutionalization of science in terms of specific fields, degree programs, and so forth), within which Darwin was working.

- There is intra-species variation.
- There is a struggle for existence.

_____

- If a variation conveys an advantage, however small, in terms of survival, organisms with the variation will be more likely survive than those without it, and will tend to pass on the variation to their offspring.

Over time, Darwin argued, the accumulation of variations within a species would lead to the prevalence of a new trait, and eventually, if enough new traits emerged, to the emergence of a new species (ibid., 80–96).

That members of a species differ in terms of some of their traits was well known at the time, and used by breeders and botanists to "select for" desirable traits through selective breeding. Darwin called this practice "artificial selection," as the traits in question are selected by breeders, rather than the result of "natural" causes. The second premise of his argument, that there is a struggle for existence, was also accepted by many contemporary scientists, having been argued for by Rev. (Thomas) Robert Malthus (1766–1834), a sociologist and cleric who focused on the growth of human populations (Malthus 1798). Malthus' argument for what he called "a population principle" proposed a potential geometric increase (i.e., an exponential increase) in human populations, given that any human can produce more than one offspring. In contrast, Malthus argued, in terms of the food, space, and other things required for humans to survive, the increase would only be arithmetic (i.e., limited to some constant amount). Based on these two mathematical "facts" (the resources humans need can only increase arithmetically while human population growth is exponential), Malthus argued that humans are and will consistently be engaged in "a struggle for existence."

Of course, if one extends the potential for geometric increase to all sexually reproducing organisms – which Malthus did not – then the potential for geometric increase also applies to many organisms that humans rely on for food, clothing, and so forth. But this does not counter his argument about limits of space and potential human population growth which (barring unlikely measures such as the reclamation of vast areas of land from the oceans) will not increase by very much, even if we begin to populate areas that are now relatively unpopulated. We also have no assurance that humans will not deplete the numbers of other species (for example, by over fishing, which is in fact occurring) on which we depend.

Although inspired by Malthus, Darwin's argument for a struggle for existence differed in important respects from that which Malthus offered. For one thing, Darwin extended the struggle to include all species. In addition, he writes in the *Origin* that he uses the phrase 'struggle for existence' in a "broad, metaphorical sense" to include ecological factors and not just inter- and intraspecies competition, or predator–prey relationships (Darwin 1859, 62). One of his examples to illustrate this broader sense is that of a plant at the edge of a desert struggling for water; others include cases in which one species' reproductive success depends on members of another species (for example, the reliance of mistletoe on birds and insects for successful propagation). In short, for Darwin, having been whelped, spawned, hatched, germinated, or born ensures not a long and rich life, but a lifelong struggle for survival. All living things require many things to survive. Finally, Darwin's argument that there is a struggle for existence did not rely solely on Malthus' mathematical formulation as Darwin provided many empirical examples illustrating the kinds of struggle organisms face in their efforts to survive, including examples of interspecies dependence (ibid., 63–79).

Having presented his general argument for natural selection, Darwin went on to argue that any trait deleterious to an organism's survival would necessarily be selected *against* and that natural selection *cannot* "select for" such traits. In chapter 4 of the *Origin*, Darwin makes the point this way:

> It may metaphorically be said that natural selection is daily and hourly scrutinizing, throughout the world, the slightest variations; rejecting those that are bad, preserving and adding up all that are good; silently and insensibly working, whenever and wherever opportunity offers, at the improvement of each organic being in relation to its organic and inorganic conditions of life. (ibid., 84)

It is important to be clear, as Darwin himself is, that his language here is deeply metaphorical; in the case of natural selection, there is no "agent," as there is in artificial selection (e.g., a breeder), selecting for or against traits. As Darwin's general argument makes clear, "natural selection" comes about because of intra-species variations (the source of which neither he nor his contemporaries understood), and a struggle for existence.

We are now in a position to understand why Darwin turned his attention to the sexes and proposed sexual selection as a secondary evolutionary mechanism. He was convinced that some traits, found predominantly among males of various species, could not be explained by natural selection – indeed, might even serve as counter-examples to or "falsifiers" of it. The traits in question not only didn't seem to contribute to survival but could be viewed, and were viewed by Darwin, as acting against it! The peacock's very large and colorful tail (which Darwin confessed in a letter to Asa Gray in 1860, made him sick whenever he observed it) is an apt example (Darwin 1860). The tail of a peacock is sufficiently flamboyant to attract predators and sufficiently cumbersome to hamper a peacock's ability to take flight, or run on the ground, to escape from predators. In contrast, peahens have small brown tails (see Figure 2.1). Darwin also noted the bright colors sported by the males of many other species (whose female counterparts are relatively drab), and the larger size of males relative to females in many species. He believed both were likely to act against survival by alerting and attracting predators. And he noted that antlers sported by pubescent and adult males of various species (deer, elk, and moose, among them) make navigating around trees, for example, and other obstacles, difficult, and can result in injuries to those endowed with them. Large racks of antlers also make animals more visible to predators. How, Darwin asked, could natural selection allow for, rather than select *against*, such traits?

Darwin's solution, his hypothesis of sexual selection, was nothing short of ingenious. This is in part because it was not an *ad hoc* hypothesis – which, as we noted in Chapter 1, is a hypothesis introduced only to save a theory from refutation and for which there is no independent evidence. From the outset of the *Origin*, Darwin emphasized the importance of *reproductive success*. Heritable variation is *required* for evolution and an organism's survival is necessary to but not sufficient for its success in reproducing. The goal – whether we are focusing, as Darwin was, on organisms or, in more recent accounts, on genes – is, respectively, to reproduce or replicate – to have

Figure 2.1 Peacock wooing Peahen.
Angel Traversi/EyeEm/Getty Images.

offspring or, in the case of genes, to have "a copy of oneself" in the next generation. If the traits in question (larger size, bright plumage, antlers, etc.) contribute to *reproductive success*, they pose no challenge to the role or efficacy of natural selection. As Darwin describes sexual selection in *The Descent of Man*, a work devoted to making the case for human evolution, it involves "the advantage which certain individuals have over other individuals *of the same sex and species, in exclusive relation to reproduction*" (1871, Vol. 1, 256; emphasis added).

In both the *Origin* and *The Descent*, Darwin defined sexual selection as including two processes: members of one sex (typically males) compete with one another for access to the other sex (typically females); and, except in the human case, females choose mates. In so much as traits such as large size and antlers help males compete with other males for access to females, and to the extent that flamboyant traits such as bright colors, and, in the case of peacocks, a larger and more eye-studded tail, are preferred by females, sexual selection explains the traits that initially concerned Darwin. Darwin mused that female choice of flamboyant traits might be an aesthetic preference; in more recent accounts of sexual selection, such traits are often understood as indicating superior fitness because bearing and sustaining them requires greater energy, and that perceived fitness motivates females' choices of mates as it promises greater fitness of their offspring. A reader might well wonder if the foregoing discussion of "choices made" by non-human females isn't anthropomorphic, a point to which we will return. When Darwin turns to sexual selection among humans in *The Descent*, he discusses cultural differences in standards of "attractiveness." He also maintains that, unlike sexual selection in most other species, it is men who choose women and women who adorn themselves to be attractive to men (Darwin 1871, 8).

We have not yet considered Darwin's hypotheses about sex differences in other traits, which he called "secondary sex characteristics." In the *Origin*, Darwin maintained that the behavior and psychology of males and females differ in significant ways. For example, he described males (remember that in this work he is discussing many species) as "generally eager to mate with any female" and females as either less eager and/or possibly afraid of a large male approaching them (ibid., 156–158). As we will see, in *The Descent*, Darwin would describe women as exhibiting coy-like behavior – a description that, together with that of men as "eager to mate," took hold among some evolutionary theorists including some contemporary theorists, as we discuss in later chapters, as well as feminists' critiques of hypotheses that assume them.

Darwin also assumed that the differences in mating strategies and related behaviors he noted resulted in what he and many of his contemporaries took to be males' obvious superiority in terms of some specific characteristics and abilities. Beginning in the *Origin*, Darwin maintained that males, given their need to compete for females and to provide for them and their offspring, develop stronger passions, heightened intelligence, greater strength and sensory capacities, and heightened locomotive abilities. In *The Descent*, Darwin attributed these and other characteristics to men, albeit in language specific to humans (ibid., 582). He attributed men's superior powers of "observation, reason, invention, and imagination" to selection pressures that he thought were unique to them – competition for mates, protecting and providing for women and children, and "defense of community" (ibid., 565). Darwin's assumptions and arguments about sex and sex/gender differences are obvious examples of biological determinism – the thesis that biology (in this case, traits that result from sexual selection) determines non-physical as well as physical traits. As noted, we discuss biological determinism in some detail in Chapter 3.

In many passages in which he addresses sex/gender differences, Darwin apparently just assumed, without reflection or investigation, the Victorian view (also common among Europeans) that men and women differ in many ways beyond anatomical and physiological differences related to reproduction, and that men are superior to women in all or most of the ways in which they differ. And he attributed men's superiority to women to the greater selection pressures he assumed men face. Viewed in this way, Darwin's view of men as superior to women seems to be not a well-thought-out and empirically based view, but simply a reflection of the culture within which he lived and worked. But in the following passage in *The Descent*, Darwin does seem to do more than just assume male superiority; he seems to offer an argument as to why men are superior in terms of some specific, culturally-valued, characteristics.

> The chief distinction in the intellectual powers of the two sexes is shown by man's attaining to a higher eminence, in whatever he takes up, than can women – whether requiring deep thought, reason, imagination, or merely the use of the senses and the hands.... We may also infer ... that if men are capable of a decided pre-eminence over women in many subjects, the average of mental power in men must be above that of women. (ibid., 327)

In men, Darwin argued, greater selection pressures lead to "higher mental faculties ... [and] will thus have been continually put to the test and selected during manhood." He concludes this discussion with the claim that "man has ultimately become superior to woman" (ibid., 328–329).

Finally, although Darwin acknowledged that women enjoy some of the qualities just mentioned, including intelligence, he maintained that they do so to a lesser degree and only *because* they inherit them from their fathers, as the following passage from *The Descent* illustrates:

> It is indeed fortunate that the laws of equal transmission of characters to both sexes prevail with mammals; otherwise, it is probable that man would have become as superior in mental endowment to women as the peacock is in ornamental plumage to the peahen. (ibid., 329)

Many feminist scientists and science scholars have offered sustained critiques of the assumptions, claims, and arguments just summarized. One concern is reflected in the title of an article by biologist Ruth Hubbard, which asks "Have Only Men Evolved?" Darwin's account of sexual selection and the sex differences he associated with it, Hubbard argues, suggest that females, including women, have enjoyed "a free evolutionary ride" on the backs of their male conspecifics (Hubbard 1983). Other feminists have offered related arguments and counter examples.

For example, feminists have criticized Darwin's assumption that a woman's survival and that of her offspring *require* the assistance of one or more men who protect and provide for them as reflecting an acceptance of unwarranted gender stereotypes. And feminists who study animal behavior provide examples involving other species (some of these examples we consider in later chapters), to argue that Darwin's claim that in most species females are dependent on males is unwarranted. Many feminists also view Darwin's arguments for sex/gender differences in "secondary characteristics," particularly his attributions of superiority to males and men in terms of intelligence and reason, as not nearly as related (if related at all) to his concerns about apparent counter-examples to natural selection (those involving size, plumage, and the like among males), as they were related to then current sociopolitical views about sex/gender (e.g., Hrdy 1986 and Hubbard 1983). They argue that the characteristics Darwin attributed to women and men, and females and males generally, reflected longstanding, deeply held, and at the time untested, assumptions about the sexes. Later in this chapter, as earlier noted, we consider arguments that scientific observations are in part shaped by background assumptions and/or accepted theories, and note the relevance of this issue to Darwin's theorizing, particularly how gender stereotypes contributed to his observations of males and females, including men and women.

Some feminists also argue that, contrary to the way he is generally portrayed by historians of science, there are several respects in which Darwin was decidedly not "swimming upstream" – that is, he was not critically taking on prevailing sociopolitical *or* scientific views. He was, as we have seen, assuming the gender stereotypes of his day (e.g., Hubbard 1983). In addition, some feminists and others point out that Darwin's model of natural selection – which involves waves of competition for scarce resources, and "winners and losers" – paralleled then current arguments for capitalism. So, too, Darwin assumed then current sociopolitical beliefs about race differences. Although an Abolitionist, he appealed to differences between "the races" in brain size and intelligence to further his argument in *The Descent* that variations in brains and intelligence could have led to the evolution of the human brain. (Explaining the human brain or mind was one of the largest hurdles he faced in making the case for human evolution.)

Many (if not most) Darwin scholars recognize that the claims Darwin makes about sex/gender, sex, and racial differences that feminists criticize are in fact unsupported assumptions, assumptions that were characteristic of Victorian England. But there are examples of Darwin going against other prevailing contemporary views, including those of contemporary scientists. A few examples are instructive. Darwin rejected the view of "fitness" as an "absolute" and stable property of organisms. Taken as a whole, his arguments concerning natural selection make it clear that he understood fitness to be an inherently *contingent* property. As he noted, changes in an organism's environment can render a previously fitness-enhancing trait deleterious. Whatever parallels exist between models of capitalism and Darwin's account of natural selection, this is a major difference between the two because many of capitalism's advocates assumed or argued that the "fitness" that determined degrees of economic success is an innate and stable property of individuals. We have also seen that Darwin challenged the view of the struggle for existence as limited to intra and interspecies competition, and predator-prey relationships (a view often described as "nature red in tooth and claw"), and developed a more nuanced view of this struggle.

That Darwin did challenge the views just noted does show that he did not accept all the views characteristic of Victorian England, but it does *not* show that feminists' critiques of his acceptance of then prevalent views about sex/gender are unwarranted. Given that he did challenge some views held by his contemporaries, feminist critics can ask if it is not reasonable to maintain that Darwin could and should have challenged many of the assumptions that informed his reasoning about the sexes and sex/genders. One obvious rejoinder is to point out that, in fact, it is not reasonable to demand of any scientist (or indeed anyone) that they identify and scrutinize all the assumptions they carry with them. We all work with the evidence available to us and quite often within the constraints brought about by our upbringing, our training, and accepted theories.

So, it is reasonable to explore in more detail the nature of the evidence about sex/gender differences that was available to Darwin, particularly as that evidence relates to his assumptions about women's dependency on men. (We will defer discussion of his descriptions of females and males of other species to later chapters.) Theorizing, as he was, in the social and political context of mid to late nineteenth century England, Darwin's observations of women's dependence on men – *as they involved women of his own race and social class* – were well in keeping with what the social and legal restrictions placed on such women resulted in. Access to education, property rights, and possibilities for anything like a "career" of the kind men of his race and social class enjoyed, were virtually closed to many of these women. As a result, many women of Darwin's time, culture, social class, and race *did* depend on males for protection and resources for themselves and their offspring. Given this context, Darwin's argument, cited above, that men undergo greater selection pressures and that they excel in many contexts (in which women had little if any opportunities to participate), did enjoy a degree of empirical support. However, as the emphasis placed on women of his own class and race indicates, Darwin's view of women was severely limited. Working class women, female servants, and female slaves of the time were not viewed as "frail" or dependent on men, as Sojourner Truth's "Ain't I a Woman?" makes clear (Truth 1851).

One can also argue that Darwin should have considered the possibility that social institutions and norms that severely limited girls' and women's activities and opportunities are what actually explained their dependence on men, and their apparent "inferiority" in terms of intellect, leadership, and passion. And this point is brought home by the fact that two of Darwin's most well-known contemporary scientists, John Stuart Mill and Alfred Wallace (the latter, as we noted, independently discovered natural selection), did advocate sociological rather than biological explanations for the apparent differences between women and men, and viewed the relevant sociological structures as unjust. In Darwin's defense, and this is not meant or intended as a flippant comment, it is quite likely that Darwin was not reading the same books, or encountering or engaging in the same arguments as were Mill and Wallace. What we know about what Darwin was reading and about his correspondence, which does include exchanges with women affiliated with feminist causes, does not indicate that he was engaging with feminist ideas. Mill, on the other hand, is thought to have been at least partly educated about feminist ideas by his wife, Harriet Taylor, who was a feminist. In addition, Mill studied and wrote about The Seneca Falls Convention in America in 1848 in which feminist Mary Wollstonecraft's arguments figured prominently.

We should also note that Darwin was certainly challenging prevailing views of many of his scientific and social contemporaries by arguing for "female choice" in nonhuman species. Some opposition to female choice was due to the assumption that females are incapable of contributing to the directions of evolutionary change. But Wallace objected to sexual selection (which included female choice) because he thought the hypothesis did not meet then current standards for what counts as a genuinely scientific theory. The then current view (discussed in *The Principles of Geology* authored by the geologist Charles Lyell) was that a genuine scientific theory would propose only *one* causal mechanism to explain the phenomena of its domain (Lyell 1830). In proposing a second evolutionary mechanism, Darwin violated this requirement – and, Wallace believed, compromised the theory of evolution by natural selection. He did not, however, convince Darwin to abandon sexual selection.

Another significant barrier to accepting female choice in non-human species was the view, common among Darwin's scientific and social contemporaries, that humans are inherently "different from animals." Only humans, it was thought, are endowed with intellectual capacities and the ability to engage in behaviors that aren't solely instinctual. So, the very idea of a peahen or other non-human female (or male animal, for that matter) being capable of anything like "choice" was simply incompatible with accepted views about characteristics taken to be unique to humans. Although Darwin would argue in *The Descent*, again in service of an argument for human evolution, that many other animals had mental capacities that differed *in degree but not in kind* from those of humans, this was not generally accepted.

It is also interesting, although we can do no more than mention it here, that prominent female authors who were contemporaries of Darwin, including George Eliot, were intrigued by Darwin's account of sexual selection and this is reflected in their novels (Beer 2009; Ruse 2017). Like many Victorians, Eliot and other female literary figures of the time were fascinated by science, particularly in terms of the insights it could provide into human nature. In Eliot's case, Darwin's account of sexual selection in the *Origin* had a strong influence in her characterizations of male–female relationships. For example, in *Daniel Deronda*, Elliot's characters view a good marriage partner for a woman to be a successfully financial man who can provide for her. Yet, even in that novel, commentators note that she also breaks with Darwin by attributing "female choice" to women. Her female characters are at least as active as their male counterparts in pursuing mates. And in *The Mill on the Floss*, Eliot turns the tables, so to speak, on Darwin's assumptions about sex/gender differences; her protagonists are a brother and a sister, and it is the sister who is the cleverer of the two and more interested in actively engaging members of "the opposite sex."

Perhaps as interesting, Darwin corresponded with women, such as Elliot, who held feminist views (albeit, they did not discuss those views with one another) and he encouraged women to engage in science. Of the 15,000 pieces of correspondence attained and archived by the Darwin Correspondence Project based at Cambridge University Library, around 600 are letters Darwin received from and wrote to about 150 women. Indeed, the work done by the "Darwin and Gender" project, which is part of the Darwin Correspondence project, reveals that "he relied on a range of women correspondents for help with some of his most serious work." Indeed, his correspondence reveals that "he also helped many to progress their own scientific careers" (Darwin and Gender project 2013). For example, in 1872 he wrote to the female scientist Mary Treat concerning her work on butterflies. She should, he wrote, "publish the result in some well-known scientific journal" (ibid.). As Phillippa Hardman, a research associate on the Darwin and Gender project, argues, "Darwin was no feminist, but our research has shown that his views on gender were a lot more complex than has been acknowledged in the past" (ibid.). Juxtaposing what he said of women in *The Descent* with his correspondence with women about science, the most reasonable conclusion might well be that which Hardman draws. At the same time, what Darwin wrote publicly about women cannot be ignored; nor can we ignore its influence on scientists and policy makers, or how they worked to severely limit women's opportunities for many generations.

We have considered several reasons to view Darwin's theorizing as good science, while granting that his assumptions about sex/gender were unquestioned and untested, something we can see but it is unclear that we should expect Darwin to have seen. Indeed, by studying Darwin's unquestioned assumptions and their consequences for his theory, scientists, and historians and philosophers of science, and not just feminists in these fields, may better understand the relationships that can obtain between *good* science and its external and internal contexts. At the same time, we also noted that there are feminists whose views about Darwin are different from those just outlined. We return to these views of Darwin's science in the next section. Such differences among feminists, and we will encounter more in the chapters ahead, illustrate that feminist theorizing within and about biology is not monolithic.

More Philosophical Issues

Our discussion of Darwin's arguments for sexual selection, and for sex/ gender and sex differences, involve important philosophical issues that historians and philosophers of science, and many scientists (including but by no means limited to feminist scientists), explore. Two such issues are addressed in this section and two others are addressed in the conclusion of this chapter. We begin by discussing "Contextualism," one of several epistemological approaches to studying science, and very briefly contrast it with two other general approaches: Objectivism, which we discussed in Chapter 1, and Postmodernism.

Contextualism

Broadly defined, a scientist or science scholar who approaches her or his study of science from the perspective of "Contextualism" believes, based on evidence they take to support this view, that the sciences have reflected and will continue to reflect, to varying degrees and in various ways, the historical and cultural contexts, both internal and external to science communities, within which they are undertaken. Moreover, many view this historical and contextual embeddedness as not only unavoidable, but also as unproblematic *provided that evidence is given its due*. We will see as we continue to consider feminists' engagements with biology, that most feminists assume one or another version of Contextualism. In this text, I refer to science in which social and historical context is of consequence, *and* evidence is given its due, as "evidentially normal science." Of course, there are cases in which these contexts are consequential, but evidence is not or was not given its due, and they are decidedly problematic; they suggest a lack of scientific "integrity" and they are accurately described as "bad science." As was clear in the foregoing discussion of feminist approaches to and critiques of Darwin, and will be clear in subsequent chapters, appeals to evidence are a central feature of feminist analyses of biology.

In Chapter 1, we considered arguments offered in the twentieth century by philosophers of science for the thesis of underdetermination. As it relates to the issues that here concern us, this is the thesis that all the evidence we have and can possibly have for any specific theory, and indeed for all our theories, cannot prove these theories. There is, in Quine's words, "empirical slack" between all the evidence available to us and the theories we propose and accept (Quine 1966, 241). Hence, contextualists argue, things like background assumptions, scientific and, in some cases, assumptions reflecting social beliefs, mediate the relationship between our sensory experiences and our theories. But the thesis of underdetermination does not exhaust the arguments for Contextualism.

The history of science, advocates of Contextualism argue, provides strong empirical evidence supporting this view of science. We have noted that Darwin's arguments for sexual selection, and the sex/gender differences he attributed to it, are recognized to reflect then current social and scientific views about sex/gender differences (among other things). As we have seen, feminist scientists and science scholars are divided in terms of how they view the relative weight and implications of these relationships. Some argue that neither Darwin nor most of his scientific contemporaries did pay sufficient attention to available evidence that challenged their assumptions about sex/gender differences. Social and political beliefs and values, rather than "good science," they conclude, prevailed. An alternative view, explored in the previous section, is that Darwin gave the evidence *most readily available to him* its due, and that there isn't evidence that his arguments for sex/gender differences were motivated by an interest in maintaining male dominance. Each position has its problems. We earlier noted problems facing the first when we considered that Darwin's assumptions about sex differences were well in keeping with his own limited experiences and observations, as well as with views commonly held by those in his scientific and social circles. That is, from the perspective of Contextualism, it is not clear that Darwin should "have known better."

But this position also has a problem. As we noted, there were, among his scientific contemporaries, some who did recognize the unquestioned assumptions about sex/gender that informed his science, did subject them to critical scrutiny, and argued that they were not warranted by available evidence. Mill cited the lack of evidence for biological, rather than sociological, causes, for such differences in his "The Subjection of Women," published two years before Darwin's *Descent* was published (Mill 1869), noting that given that women were not afforded the same opportunities, in say education, as men, the hypothesis that they were intellectually inferior had never been tested let alone confirmed. But although disagreements remain about how to evaluate Darwin's reasoning about sex/gender differences, note that each of the positions we've considered approaches science through the lens of Contextualism. Each cites social and scientific contexts as having a significant impact on Darwin's reasoning.

One way to provide more content to Contextualism this early in our discussion is to juxtapose it to two other epistemological approaches. As most feminist scientists and science scholars embrace some version of Contextualism, our discussions of the alternatives are brief; their purpose is to illuminate arguments for and aspects of the contextualist approaches we will encounter in the chapters to come.

We begin with what is commonly referred to as "Objectivism," which was introduced in Chapter 1. Like Contextualism, there are versions of Objectivism so we focus on aspects that they have in common. We encountered some of these features in Sheldon Glashow's statement summarizing what he argued to be scientists' commitments. These include metaphysical commitments, which involve beliefs that nature contains (or does not contain) one or more kinds of objects, events, and/or processes. In Glashow's case, he cites a commitment to the view that there are eternal, objective, and context-free laws of nature. The emphasis Glashow places on "laws" reflects the long-standing and deep-seated interest in physics in finding laws of the kind he describes. The more general metaphysical commitment that is compatible with this, but also applicable to biology and other sciences that do not include laws, is that there is a world *completely independent of* our concepts, theories, and interests; and that *it is features of this world* that scientists seek to identify. Metaphysical commitments, as understood in the philosophy of science, are beliefs about "the nature of nature" (e.g., that it contains only matter and energy) or some aspect thereof (e.g., that minds are or are not nonphysical objects).

Glashow's tenets also include epistemological commitments. Epistemological commitments concern knowledge: its nature, its sources, and its limits. Glashow's tenets include that humans have the cognitive and sensory capacities to know features of nature and that such knowledge is intersubjective (indeed, he states that even "aliens" would arrive at the same knowledge as we). Another of Glashow's epistemological commitments is that "what we call physical science" has already discovered at least some eternal, universal, socially neutral laws of nature.

There are various arguments for Objectivism. Here, as in Chapter 1, we note the argument that many find most compelling. Objectivists maintain that science's success in explaining, manipulating, and predicting aspects of nature would be nothing short of *miraculous* if there were not a world that exists independently of specific contexts, concepts, and theories, *and* if science has not succeeded in discovering truths about it. Objectivists hold that Contextualism is incapable of explaining science's success – something most advocates of Contextualism deny.

In sharp contrast to Contextualism and Objectivism, Postmodernist approaches to science, including those advocated by some feminists, view scientific theories to be socially constructed "myths." Accordingly, they should be read as texts, as narratives, in order "to reveal [their] cultural meaning" (Sperling 1991, 25). So, too, "scientific facts" are taken to be socially constructed. Moreover, some postmodernists maintain that science itself, to paraphrase the title of an article by primatologist Donna Haraway, "is politics by other means" (Haraway 1984). As philosopher Elizabeth Anderson describes the rationale of Postmodernism, arguments for it have their roots in "ideas about language and thought. "[Postmodernists] claim that (what we think of as) reality is 'discursively constructed'" (Anderson 2015). Anderson, it is important to note, is not a postmodernist, but her account of Postmodernism is more accessible than that given by many postmodernists; the latter are, frankly, very difficult to decipher. Given that feminist postmodernist approaches to science are relatively rare, we need not engage in a prolonged analysis of the arguments for Postmodernism. But we should note just *why* they are rare.

Briefly put, many feminist scientists and science scholars view postmodernist approaches as undermining the critical and constructive contributions feminists have made to science. If all theories are "myths," so too are those feminists develop; if all facts are "socially constructed," so too are those facts feminists point to as *evidence* in their critiques and constructive alternatives. And in more general terms, as primatologist Susan Sperling argues, "Postmodernism leaves untouched the question of the relative worth of epistemologies of [science – that is, theories about science], because postmodernists tend to view all epistemologies as equally mythic social constructions" (Sperling 1991, 26).

We have noted that, in general, feminist scientists and science scholars adopt one or another version of Contextualism in studying specific cases and fields in science. We have mentioned two general lines of argument for their doing so: appeals to the thesis of underdetermination (to which we return below), and appeals to numerous historical examples in which features of then current scientific and social contexts had a role, to some degree or other, in shaping scientists' reasoning and conclusions. Our discussion of Darwin has already introduced one case involving the role of historical, social, and scientific contexts. In Chapter 1, we also considered three cases in which background assumptions (scientific in two cases, and sociocultural as well as scientific in Aristotle's case) figured in scientific reasoning. We also encountered Carl Hempel's arguments that the test of any hypothesis makes use of such assumptions. The role of background assumptions in scientific theorizing, a role we will see that is argued to extend beyond tests of hypotheses and theories, is also understood to support Contextualism. Whether the assumptions at work are scientific, social, or a combination thereof, they reflect aspects of the contexts within which science is being undertaken.

Contextualism's roots in the philosophy of science (which may differ from its roots in history of science) date at least to arguments offered in the 1930s by Otto Neurath who was a prominent member of the Vienna Circle. This was a group that included philosophers with scientific training and scientists who had philosophical interests that met in Vienna in the 1920s and 1930s and sought to clarify the nature of scientific knowledge and of logic and mathematics. Neurath rejected foundationalism – the view that there are "pre-theoretical" statements that are in some sense obvious (not in need of support or evidence for them) and such that all of science is based on them. (A common form of foundationalism took these statements to be about sense data, that is, about what we directly perceive, e.g. '"red, here, now.") Neurath viewed science as an evolving body of interlinked observational and theoretical claims put forward in specific historical, social, and political contexts, all of which are subject to revision and whose justification is proportional to their success in making sense of the world and our interactions with it.

In the 1960s, the philosopher W.V. Quine also advanced arguments for Contextualism and frequently cited a now famous metaphor used by Neurath. In Quine's words: "Neurath has likened science to a boat, which if we are to rebuild it, we must rebuild plank by plank while staying afloat in it" (Quine 1960, 3). Quine also offered the following description of his own version of Contextualism: "Our one serious conceptual scheme," he argued, "is the inclusive, evolving one of science, which we inherit and, in our several small ways, help to improve" (Quine 1966, 239).

Quine also articulated what is arguably the clearest statement of the thesis of underdetermination as this thesis relates to the issues that will concern us: that all the evidence we have and could possibly have for any specific theory, and indeed for all our theories (including those of "common sense" and philosophy), does not entail (i.e., prove) these theories. As we earlier noted, there is, in Quine's words, "empirical slack" between all the evidence available to us and the theories we propose and accept. Hence, contextualists argue, factors such as background assumptions, scientific and/or social, mediate the relationship between our sensory experiences and hypotheses and theories.

Another thesis that emerged in mid-twentieth century philosophy of science is that observations, in science and everyday life, are in part shaped by current theories. This thesis, which we consider following this discussion of Contextualism, is also understood as supporting Contextualism. But we can anticipate that discussion by remembering Darwin's "observations" of specific and significant differences between women and men, and how they reflected then current social beliefs and policies.

As earlier noted, feminists have offered versions of Contextualism. What is common to them, and what distinguishes them from some non-feminist versions of Contextualism, is that feminists maintain that the sociopolitical contexts, including accepted social values, within which science is done, can and often have informed scientific questions and hypotheses. We have encountered gender stereotypes in Darwin's theorizing, and we will consider feminists' arguments for the role of gendered and evaluatively -thick metaphors when we turn to Parental Investment Theory and Human Sociobiology in Chapter 3; both, feminists argue, reflect social beliefs and values.

One feminist contextualist model of scientific reasoning was put forward in an article co-authored by philosopher Helen E. Longino and biologist Ruth Doell (Longino and Doell 1983). Longino's and Doell's analysis focuses on three aspects of scientific theorizing that make it possible for *androcentrism* (again, this is understood as "male-centeredness") to find its way into *good* science: the determination of facts, the determination of which facts count as evidence, and the background assumptions that result in a fact being viewed as evidence for a hypothesis (Longino and Doell 1983).

Facts, Longino and Doell argue, are the "least defeasible" aspect of scientific reasoning. That is, they are the parts of science that are least likely to be in dispute, but even they are not as straightforward as might be assumed. Nor is the status of a fact *as evidence* for a given hypothesis obvious or "self-announcing." Longino and Doell offer several arguments to support these claims. First, they point out, we do not recognize or take account of all "facts"; for example, we are not normally cognizant of each leaf on every tree we pass on a morning walk. In addition, an object or situation (i.e., something that can constitute "a fact") can be correctly described in more than one way. For example, a specific situation two of us observe might be correctly described either as, "There is a tattered backpack on the couch," which might be how I describe it; or as, "There is a red, full backpack on the couch," which might be how you describe it. Further, two of us may take the fact as we describe it as evidence for the same hypothesis – for example, "Mary is home" – but for different reasons. I may take the backpack as evidence that Mary is home because Mary always leaves her backpack on the couch when she returns home; you may take the presence of a red and full backpack as evidence that Mary is home because, wherever she might leave it, this is what her backpack looks like. In other words, we can bring different background assumptions to bear in coming to take a fact to be evidence for a given hypothesis.

Finally, we might agree on our description of a fact (for example, in contrast to the example cited above, we might both accept the fact "There is a tattered backpack on the couch"). But I may take this fact as evidence for the hypothesis "John has come to visit" and you may take it as evidence for the hypotheses "Mary is home." This will happen if the two of us bring different background assumptions that render the fact or facts we both accept as evidence for *different hypotheses*. In the case at hand, I may know that Mary is in fact out of town and that John's backpack looks just like hers. You may not know that Mary is out of town or that John has a backpack that looks just like hers.

Given that we notice or pay attention to only some facts, that the same empirical situation can be correctly described in more than one way, that facts are taken to be evidence only in relation to some hypothesis, and that background assumptions help to mediate the relationship between the facts recognized as evidence and the hypothesis for which they are taken to be evidence, Longino and Doell argue, there are three ways in which androcentrism can make its way into scientific reasoning. The accepted *facts* might themselves be specified in androcentric terms, and they might be taken *as evidence* that, given one or more *background assumptions* that may be androcentric, support a *hypothesis* that is itself androcentric.

Longino and Doell use this model to analyze several cases in which feminists have argued that androcentrism informed scientific research. Some of these we consider in subsequent chapters. Here we will focus on one theory they considered that was of interest to feminists in the 1970s and 1980s – a theory about human evolution called "Man, the Hunter" theory. Feminists argued that the theory was fundamentally androcentric. Longino and Doell use their model to analyze how androcentrism informed the reasoning to, and content of, the theory, and take their analysis to illustrate that androcentrism can inform *good science* – which they take "Man, the hunter" theory to be despite the problems they and other feminists cite. It was good science in the sense discussed in Chapter 1.

"Man, the hunter" theory was developed by physical anthropologists in the 1960s. It attributed the evolution of the human brain, the emergence of social organization, and the emergence of language, to hunting engaged in by our male ancestors. Indeed, the emergence of the hunting of large animals was called "the hunting adaptation." Building from anthropological studies of contemporary hunter-gatherer groups, its developers assumed that the gendered divisions of labor described in ethnographic studies of these populations were also characteristic of ancestral populations. These ethnographic studies reported that men undertake hunting and often travel long distances to do so; and that women are primarily responsible for gathering plants close to home and taking care of offspring. Hunting was credited with providing the most important sustenance for the group, and men were also reported to be dominant over women. (We will see that, although this view of sexual divisions of labor among our ancestors was eventually abandoned by anthropologists when evidence was discovered that severely challenged it, Human Sociobiologists continued to assume it in the 1980s and contemporary Human Evolutionary Psychologists still assume it. For both programs, it has served as the basis for many of the evolutionary explanations they offer about alleged sex/gender differences.)

In the 1970s, before "Man, the hunter" theory was abandoned, there were things about which its advocates and feminists critical of its androcentrism, agreed. Chipped stones were found near ancient human living sites, as were the fossilized carcasses of large animals that had been butchered. "Man, the hunter" theorists took the chipped stones to be evidence of men making tools to use in hunting and the butchered remains as evidence of "the hunting adaptation." Feminists critical of the theory claimed it was flawed in several respects. They argued that the ethnographic accounts of contemporary hunter-gatherer groups, produced in the main by male researchers in the 1950s, were both androcentric and ethnocentric, imposing assumptions about men's dominance and gendered divisions of labor that reflected contemporary Western societies. Subsequent studies that female anthropologists undertook (and in which they, unlike earlier male anthropologists, were allowed to interact with and interview women) challenged earlier reports that men are dominant in these groups; the newer studies also suggested that the hunting of large animals is largely symbolic and enables men to bond with each other; that in most such groups, the gathering women engage in is a highly skilled activity and is the major source of food; and that childcare is shared among members of the group (Slocum 1975).

Thus, feminists argued, the reconstruction of ancestral hunter-gatherer groups was based on flawed, in the sense of androcentric and ethnocentric, accounts of contemporary hunter-gatherer groups. The chipped stones, they argued, could just as well be evidence that women were making tools for gathering plants and preparing food; and evidence was emerging that the large animals that human ancestors butchered may not have been killed by human hunters but rather died of other causes and were foraged by these ancestors. Using these several lines of evidence and argument, feminists maintained that there was no evidence that our female ancestors did not contribute as much as their male counterparts to the survival and cohesion of their groups, or to human brain evolution.

Here is a case, Longino and Doell argue, in which the same fact (chipped stones) is taken to be evidence for conflicting hypotheses, with background assumptions mediating the relationship between the hypotheses and the fact taken to be evidence in support of them. In addition, feminists argued, the androcentrism of the "Man, the hunter" theory and of earlier accounts of contemporary hunter-gatherer groups, and the challenges that became possible due to the entrance of women into the relevant fields, are evidence of how social contexts, both internal and external to science, can inform what is, nonetheless, good science.

As noted earlier, Longino and Doell's model for analyzing how *good science* can be informed by assumptions about sex/gender (whether these are warranted or unwarranted), is one of several developed by feminists who are contextualists. Longino went on to develop this approach in substantive and influential ways in her book, *Science as Social Knowledge: Values and Objectivity in Scientific Inquiry* (Longino 1990). She calls the model of scientific inquiry she develops "contextual empiricism." Other feminists have also developed feminist versions of Contextualism (e.g., Nelson 1990; Harding 1991; Potter 2001).

Observation (Scientific and Otherwise) as "Theory-Laden"

Another philosophical issue relevant to feminists' arguments about Darwin, and to their engagements with biology more generally, concerns the nature of *observation*, including scientific observation. Until the mid-twentieth century, many scientists and philosophers of science viewed scientific observations as direct, reliable, and intersubjective. That is, assuming scientists have normal sensory apparatus and do not impose preconceptions or biases as they observe some feature or features of nature, scientists "looking at the same phenomenon" will observe "the same thing." Observation, undertaken properly, was taken to provide direct access to observable features of nature. For many who maintained that scientific reasoning begins with observations, as well as for many who maintain that hypotheses come first and are tested by means of observation – two schools of thought that dominated the philosophy of science at least until the 1980s – observations were taken to be "independent of theory" and as constituting a firm "foundation" for scientific reasoning.

However, significant challenges emerged to these views, of which we consider two. One of the earliest was offered in the 1990s by Pierre Duhem (who was mentioned in our discussion of Contextualism). Duhem used a thought experiment to illustrate his argument that the kinds of observation physicists make require extensive background in theories of physics. He asks readers to imagine a novice visiting a lab in which a physicist is performing an experiment to determine the electrical resistance of spools of wire. Both the novice and physicist see the same objects, including spools, an electric cell, small cups of mercury, copper wire, and a mirror mounted on an iron bar. As the physicist connects some of these objects to others, the iron bar oscillates and the mirror projects a band of light on a celluloid scale. Again, both the novice and physicist observe these phenomena. Duhem then has the novice ask the physicist what he is doing.

> Will he answer "I am studying the oscillations of an iron bar which carries a mirror"? No, he will say that he is measuring the electrical resistance of the spools. If you are astonished, if you ask him what his words mean, what relation they have with the phenomena he has been observing and you have observed at the same time as he, he will answer that your question requires a long explanation and that you should take a course in electricity. (Duhem 1954, 218)

Duhem's point is, of course, that the novice will need to learn some physics to be able to observe what the experimentalist observes: the level of electrical resistance.

While Duhem stressed the role of theoretical knowledge in making scientific observations possible (particularly knowledge of theories in physics), N. R. Hanson, who had been trained in anthropology as well as philosophy, offered more general arguments for the theory-ladenness of observation. He did discuss the role of scientific knowledge and training in allowing for and, to some degree, helping shape, observations, and quoted Duhem in those arguments. But the scope of Hanson's argument for theory-ladenness was much more general. Among the factors he cited as helping to shape and make possible everyday as well as scientific observation, are background knowledge, prior experience, expectations, language, and conceptual schemes. Hanson argued that cross-cultural differences in conceptual schemes and language – including differences in color schemes and mathematical systems – lead to different observations. He also used optical illusions to illustrate that although the tiny and inverted images etched on an observer's retina when looking at an object did not change, what an observer undergoing such an experience "observed" or saw in the sense of visual experience did change. Consider his example involving the Necker cube (Figure 2.2).

Here, a viewer's perspective switches from seeing the cube "from above" to seeing it "from below," although neither the object, the light rays that etch tiny inverted images on our retinas, nor those tiny images, themselves change. Consider another of Hanson's examples involving optical illusions – a sketch which can be seen as a sketch of a young woman and a sketch of an old woman (Figure 2.3).

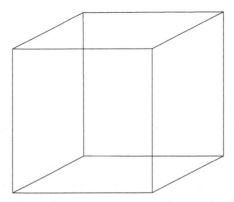

Figure 2.2 The Necker cube. "Observations on some remarkable optical phæno-mena seen in Switzerland; and on an optical phænomenon which occurs on viewing a figure of a crystal or geometrical solid," L. A. Necker, *The London and Edinburgh Philosophical Magazine and Journal of Science Series 3*, 1832.
Taylor and Francis Online, reprinted by permission of the publisher (Taylor & Francis Ltd., www.tandfonline.com).

Figure 2.3 "My wife and my mother-in-law" by W. E. Hill, Illus. in: Puck, v. 78, no. 2018 (1915 Nov. 6), p. 11.

Some people report being only being able to see the old woman, or only a young woman and often "learn to see" both an old woman and a young woman with help from others. Again, this is a case in which the object does not change, nor do the physical processes and retinal effects, but what we observe in the sense of visual experience can and often does change. Consider as well how likely it would be if someone unfamiliar with two-dimensional representations of three-dimensional objects would see, in the sense of visual experience, a woman at all.

The point of Hanson's arguments involving optical illusions was to distinguish between what he argued are two kinds of "seeing": one, the physical processes that result in retinal images that two observers looking at the same thing will share; the other, the visual experience of seeing, that can be different for observers looking at the same thing if they bring different background knowledge, prior experiences, concepts, and/or expectations to bear – an argument, we will see, that he would go on to develop.

We have noted that both Duhem and Hanson recognize the role of scientific knowledge in helping to shape scientific observations. But Hanson also argued that conceptual schemes, embodied in everyday languages, help shape observations. Consider, for example, aspects of what might be called our Western "common-sense theory" that includes categories, concepts, and objects embodied in our language: physical objects, color schemes, number systems, cause-and-effect relationships, and so forth. These, Hanson argued, help shape what we "see" or observe. If, for example, our number system was simpler than it currently is, if like a culture he encountered we counted "1, 2, 3, many," we would not observe (in the sense of visually experiencing) a pile of six pebbles. We would observe a pile of many pebbles given that "six" would not be one of our concepts.

Quine argued that every day, as well as scientific, observations are not only shaped in part by conceptual schemes – but are only possible because of them. Language, he argued, is required for the acquisition of conceptual schemes, and such schemes are required for knowledge and recoverable observations (Quine 1960). In a delightful entry in his book, *Quiddities: An Intermittently Philosophical Dictionary*, his entry "Things" begins this way:

> William James pictured the baby's senses as first assailed by a "blooming, buzzing confusion." In the fullness of time, a sorting out sets in. "Hello," he has the infant wordlessly noting, "Thingumabob again!"

Thingumabob: a rattle, perhaps, or a bottle, a ball, a towel, a mother? Or it may only have been sunshine, a cool breeze, a snatch of maternal baby talk; they are all on equal footing on first acquaintance. Later we come to recognize corporeal things as the substantial foundation of nature. The very word 'thing' connotes bodies first and foremost, and it takes some effort to appreciate that the corporeal sense of 'thing' is peculiarly sophisticated.

... Seeing a body again means to us that it or we or our glance has returned from a roundtrip in the course of which the body was out of sight. This is what is sophisticated about the corporeal of 'thing'. The baby's wordless 'again' is innocent of that. (Quine 1987, 204)

For Quine, seeing a body "again," indicates one has caught on to the notion of objects enduring through time, very much a theoretical notion.

As a final example of the relationship between scientific theories and scientific observations, consider Thomas A. Kuhn's account of the differences between what Western scientists observed prior to and following the acceptance of the Copernican hypothesis in terms of changes in the heavens. Kuhn cites these changes as evidence that theories (or what he called "paradigms") accepted by scientists help shape what they observe. Recall that, according to the Geocentric theory, change does not occur in the superlunar sphere (the sphere that includes the moon and everything beyond) but only in the sublunar sphere.

Can it conceivably be an accident ... that Western astronomers first saw change in the previously immutable heavens during the half century after Copernicus' paradigm was first proposed? The Chinese, whose cosmological beliefs did not preclude celestial change, had recorded the appearance of many new stars at a much earlier date. Also, even without the aid of a telescope, the Chinese had systematically recorded the appearance of sunspots centuries before they were seen by Galileo and his contemporaries. (Kuhn 1962, 116)

Given more recent research in neuroscience we now have a much richer (if still incomplete) understanding of the physical processes at work in vision and other sensory experience. But the philosophical issues about the role of "theories" (broadly construed to include scientific as well as common-sense theories), expectations, prior experience, and conceptual schemes in shaping observations remain relevant to our understandings of science. Neither Duhem, Hanson, Quine, Kuhn, nor most philosophers, scientists, and historians who recognize that observations are in part shaped by theories and other factors, *deny that there is a world that also shapes and constrains* what we can and do observe. Indeed, Hanson argued that, *it is only because there is such a world* that it is interesting that two observers looking, so to speak, "at the same thing" might not observe, i.e., visually experience, the same thing.

In an obvious sense, arguments that observations are theory-laden are one type of evidence taken to support Contextualism. Consider, for example, how differently Darwin and Mill perceived women's and men's abilities. Consider, as well, how differently groups of anthropologists working during different decades and with different background assumptions, perceived behaviors, and forms of social organization in contemporary hunter-gatherer groups.

As we will see in the chapters that follow, feminists argue that androcentrism and gender stereotypes have had an impact on scientific observations of the behaviors and forms of social organization of many species, including our own. We begin to introduce such arguments in some detail in Chapter 3, and continue to explore them in our discussions of specific research and hypotheses that occur in later chapters.

3   Sexual Selection

Parental Investment Theory and Human Sociobiology

Sexual Selection Theory after Darwin

We've seen that Darwin's hypothesis of sexual selection was controversial among many of his contemporaries, including his supporters. The acceptance of sexual selection took longer than did that of natural selection, and reemerged as a serious hypothesis in the early and mid-twentieth century following the emergence of genetics and of what came to be called "The Modern Synthesis." The synthesis brought together Mendelian genetics, Darwin's theory of evolution by natural selection, and other biological sciences. In this chapter, we turn to developments in sexual selection theory in evolutionary theorizing in the twentieth century, arguably the most important being the emergence and widespread acceptance of Parental Investment Theory. We discuss this theory, and feminists' critiques of it. We also discuss the emergence of Human Sociobiology in the 1970s, and the role of sexual selection theory and Parental Investment Theory in the explanations it offers of purported sex/gender differences.

Philosophical Issues

We consider feminists' arguments that unwarranted background assumptions about sex/gender shaped the account of the ancestral environment in which early humans evolved that Human Sociobiology assumes and uses to explain what it purports to be current sex/gender differences in behavior and temperament.

Following our discussions of Parental Investment Theory and Human Sociobiology, we consider two additional philosophical issues in some detail. One is biological determinism: a methodological and metaphysical approach that maintains that many features of human psychology and specific human behaviors are rooted in our biology. For parental investment theorists, purported sex and sex/gender differences in mating and parenting strategies are determined by gametic dimorphism – differences in the size and number of eggs and sperm. For human sociobiologists, although gametic dimorphism is relevant to the purported sex and sex/gender differences they propose, human social behavior and forms of social organization are determined by genes "selected for" during the Pleistocene. Darwin's arguments that different selection pressures on males and females, including evolving humans, caused sex and sex/gender differences in what he termed "secondary sex characteristics," are also biologically determinist explanations. We consider the critiques feminist offer of biological/genetic determinism in relation to the theories considered in this chapter. As will become clear, they are also relevant to hypotheses we consider in subsequent chapters.

The other philosophical issue we consider in some detail involves the role of what feminists refer to as "gendered metaphors" and "evaluatively-thick concepts" in Parental Investment Theory and Human Sociobiology. "Gendered metaphors," as feminists describe them, attribute characteristics associated with masculinity and femininity to biological entities that are not sexed. "Evaluatively-thick concepts" carry normative as well as empirical context; accordingly, feminists argue, such concepts are not "value free." In subsequent chapters, we consider feminists' arguments that gender stereotypes, "gendered metaphors," and "evaluatively-thick concepts" inform research in other biological fields.

Parental Investment Theory

The origins of Parental Investment Theory lie in 64 experiments on fruit flies (*Drosophila melanogaster*) undertaken in 1948 by geneticist Angus John Bateman. Using phenotypic traits to determine parentage (DNA testing not being available at the time), Bateman reported a greater variance among males in terms of reproductive success than among females. Appealing to this and other of his experimental results, Bateman proposed that three principles apply to many if not all sexually reproducing species: 1) males show greater variance in numbers of offspring than do females; 2) males show greater variance in the number of sexual partners they engage; and 3) there is a stronger relationship between reproductive success and multiple partners in the case of males than there is for females (Bateman 1948). Bateman took what he found to be a greater variance in male reproductive success to entail "a nearly universal dichotomy in the sexual nature of the male and female, a combination of an undiscriminating eagerness in the males and a discriminating passivity in the females," including human males and females (Bateman 1948, 365).

In 1972, evolutionary theorist Robert. L. Trivers, citing Bateman's article as his "key reference," introduced "Parental Investment Theory" (hereafter, PIT). Trivers proposed that the sex that "invests most" in offspring will be more selective in choosing a mate and that the sex that invests least will be less selective (Trivers 1972). He defined "parental investment" as "*any investment by the parent in an individual offspring that increases the offspring's chance of surviving (and hence reproductive success) at the cost of the parent's ability to invest in other offspring*" (ibid., 139; emphasis in original). Believing that it is possible to classify species in terms of "relative parental investment," Trivers argued that in the majority of species, "the male's only contribution to the survival of his offspring is his sex cells." In contrast, Trivers continued, in such species "female contributions clearly exceed males' and by a large ration" (ibid., 141). Gametic dimorphism – differences in the size and number of sperm and eggs – itself, Trivers argued, contributes to the sexually differentiated mating strategies Darwin proposed.

> Given the initial imbalance in investment the male may maximize his chances of leaving surviving offspring by copulating with and abandoning many females, some of whom, alone or with the aid of others, will raise his offspring. (ibid., 145)

This differential investment, Trivers argued, also explains the asymmetry in breeding potential that Bateman found and predicts the differences in reproductive strategies he and Darwin proposed. Although Trivers cited species that are exceptions to this generalization, humans are not among them.

> In the human species ... a copulation costing the male virtually nothing may trigger a nine-month investment by the female that is not trivial, followed, if she wishes, by a fifteen-year investment in the offspring during this period. Although the male can contribute parental care during this period, he need not necessarily do so. After a nine-month pregnancy, a female is more or less free to terminate her investment at any moment but doing so wastes her investment up until then. (ibid., 146)

So, on Trivers' view, it is greater female investment, in sex cells and in activities related to parenting (e.g., gestation and egg-sitting) that explain the differences between the sexes in mating strategies in many species (although Trivers is clear, not all species) that Darwin described and Bateman predicted. Based on greater female "investment" in offspring, Trivers referred to females as a "limited resource" in species such as ours, given gestation, lactation, and other aspects of female investment in offspring and given that females, including human females, choose mates. Unlike Darwin, Trivers and others who advocate PIT do include women among the females who choose mates. They claim that differences in parental investment explain the "greater eagerness" to mate that Darwin attributed to males, including human males; and the "coyness" Darwin attributed to females, including human females.

PIT became a mainstay of evolutionary accounts of sexual selection, even though, as we will see, it was not often tested by studying the behaviors and strategies of females and males in specific species until some decades later. At the same time, as we will see later in this chapter, and when we turn to primatology in the next, field and laboratory scientists routinely reported observations of males and females of many species whose behaviors were taken to be compatible with the theory's predictions. Many of these observations were later challenged when developments – some attributed to feminist researchers – led to better observational techniques and more attention to female behavior.

Feminist Critiques of Sexual Selection and Parental Investment Theory

Although long accepted, PIT has been the subject of sustained critique, including but not limited to, critiques offered by feminists since the early 1980s. Several challenges to Trivers' core assumption about the relative cheapness of sperm have emerged. These include that males produce millions of sperm for each egg a female produces; that in some species, high sperm production is correlated with shorter life-span; that males of many species produce costly accessory secretions in addition to sperm; and that in many species, high sperm production is required for successful insemination. Given these factors, critics argue that a more empirically warranted comparison of parental investment would invoke the minimum number of ejaculations required for fertilization and the cost of a single egg (e.g., Tang-Martinez 2000).

Nor, critics charge, does field research bear out PIT's predictions. The literature citing what are viewed to be counter-examples is extensive. Females of some species, as we will see, for example, in our discussion of nonhuman primates in Chapter 4, seek to mate more than once or twice, including when they are not ovulating and/or are pregnant. Field studies involving birds, marine organisms, leopards, and lions report females to be less than monogamous, and an extensive survey published in 1986 by Sarah Blaffer Hrdy (discussed in Chapter 4) detailed numerous counter-examples to the coy female hypothesis. Indeed, at that time, Hrdy reported, there were no fewer than six different models to explain how females might benefit from mating with multiple males. The models were developed in response to many reports, dating from as early as 1975, of polyandrous mating practices. In short, feminists argue, abundant counter-examples to the coy female hypothesis of PIT have been found over the past thirty years. In the concluding section of this chapter and in Chapter 4, we consider additional counter-examples to the theory's predictions about mating and parenting strategies. These are found in species in which female gametes are larger than males but the behavioral patterns are not the patterns PIT predicts.

Trivers and other advocates of PIT also take differential investment in offspring to entail that the interests of the sexes are inherently conflicting (often described as "the battle of the sexes" in publications intended for a lay public): that females are inherently interested in acquiring a mate who will help provide for offspring, i.e., the best mate in relation to this issue, and that males are inherently interested in acquiring as many mates as possible. But, critics argue, given that sexual reproduction is the primary function of courtship in sexually reproducing species, we should take a skeptical view of an inherent conflict between the sexes (e.g., Tang-Martinez 2000). As biologists Hamish Spencer and Judith Masters make this point, "The male-female communication system is subject to strong stabilizing selection"; thus, "unusual or fussy individuals" (i.e., the "coy females" PIT predicts) would be at a distinct disadvantage "either because they would be more likely to be rejected or more likely to reject suitable mates" (Spencer and Masters 1992).

Finally, critics point to cases involving circular reasoning in the application of the theory. In one well-known example, Trivers notes that in dung flies, "the male who first leaps on top of a newly arrived female copulates with her." Nonetheless, Trivers suggests, the lack of female choice in this case does not challenge PIT because it may "result from the *prima facie* case the first male establishes for his sound reproductive abilities" (Trivers 1972, 170). Spencer and Masters argue that, in this passage, the hypothesis of female choice is untestable (Spenser and Masters 1992, 296). Reflecting these and the other lines of argument just summarized, they conclude that

> Uncritical acceptance of [sexual selection's] ubiquitous occurrence (e.g., in Trivers 1972) does no service to evolutionary theory ... heritable variance in reproductive success, based on characters that are not favored by natural selection, must be demonstrable before sexual selection may be invoked unequivocally. (ibid., 301)

Here Spencer and Masters remind evolutionary theorists of what Darwin took sexual selection to explain. Recall that Darwin posited sexual selection as a mechanism to explain phenomena that he believed natural selection could *not* explain: traits that seemed to provide no advantage in terms of survival and, indeed, could work against it. Darwin did not assume that every trait has been selected for (that is, is an adaptation); but he did view the preponderance of such traits among males as calling for some sort of explanation. In other words, the traits to be explained by sexual selection were those that, if one only assumed natural selection, seemed inexplicable. Add the components of sexual selection, however, and they no longer seem so. But that they don't, does not mean that an explanation based on sexual selection only requires that we establish the utility of the "trait" in question. It has precisely the opposite effect. It sets severe limits on the traits to be so explained to *exclude* those that are merely adaptive and those that natural selection can explain; and to include *only those* for which there is evidence of heritable variation.

So, serious challenges to PIT have been raised. Richard Dawkins, who championed PIT in the first edition of *The Selfish Gene,* acknowledged, in an endnote in its second edition, that the assumption that sperm are cheap relative to eggs as an explanation of differential parental investment, is no longer credible. Dawkins remains convinced, however, that there is "a fundamental asymmetry in sex roles" concerning parenting and he uses an endnote annotating the original text to propose an alternative explanation of them. Although his explanation is beyond the scope of this discussion, it is interesting to note that his rejection of the role of gametic dimorphism in leading to sex-differentiated mating strategies is generally not noted by practitioners of Human Sociobiology or of Human Evolutionary Psychology for whom PIT is a core source of their explanations of sex/gender differences. In Chapter 8, we explore the role of PIT in Human Evolutionary Psychology. Here we turn to its role in Human Sociobiology.

Human Sociobiology

Human Sociobiology is a research program initially proposed by geneticist E. O. Wilson in *Sociobiology: The New Synthesis* published in 1975. As Wilson described it, the research program would explain animal social behavior on the basis of genes that were selected for to enhance survival or reproductive success. In other words, such social behaviors are *adaptations*, a term, which as we noted in Chapter 2, is reserved for traits that are the product of natural or sexual selection. Although only the first and last chapters of Wilson's volume discuss humans, in describing the scope of the disciplines Sociobiology would "synthesize" into branches of biology, Wilson pays special attention to disciplines that focus on humans – the social sciences and humanities. Despite stating that in the case of humans, "genes have given up much of their sovereignty," Wilson cited an extensive list of human behaviors as candidates for genetic, adaptation explanation: they included "male aggression," "war," "xenophobia," and "male dominance." Social sciences such as sociology, Wilson argued, which study such phenomena as well as other human social behavior and forms of social organization, merely describe "surface phenomena" that evolutionary biology can be expected to actually explain.

> Sociobiology is defined as the systematic study of the biological basis of all social behavior. For the present it focuses on animal societies, their population structure, castes, and communication, together with all of the physiology underlying the social adaptations. But the discipline is also concerned with *the behavior of early man and the adaptive features of organization in the more primitive [sic] contemporary human societies*. It may not be too much to say that sociology and the other social sciences, as well as the humanities, are the last branches of biology waiting to be included in the Modern Synthesis. (Wilson 1975, 4; emphasis added)

Although Wilson mentions "the behavior of early man" and what he describes as "the more primitive [sic] contemporary human societies" in this passage, his interest in synthesizing the social sciences and humanities – which, of course, often focus on contemporary humans – attests to the broader project described earlier.

This project is to explain *contemporary* human social behavior based on genes that were "selected for" by natural or sexual selection during the Pleistocene when it is thought distinctly human traits – such as symbolic thought, social identities, and rapid innovation – emerged, genes that, Human Sociologists maintain, we have inherited. (The Pleistocene spanned 1.8 to 0.01 million years ago; *Homo sapiens*, fossils and artifacts suggest, emerged during the later years of that period.) Subsequent passages from chapters 1 and 27 of Wilson's book make this more ambitious project clear. "There is a need," he maintains,

> for a discipline of anthropological genetics. In the interval before we acquire it, it should be possible to characterize the human biogram by two indirect methods. First, models can be constructed from the most elementary rules of human behavior ... The rules can be legitimately compared with the ethograms of other primate species ...
>
> The other indirect approach to anthropological genetics is through phylogenetic analysis. By comparing man with other primate species, it might be possible to identify basic primate traits that lie beneath the surface and help to determine the configuration of man's higher social behavior. (ibid., 551)

By 1978 when he published *On Human Nature* (a book intended for a general audience), Wilson and other human sociobiologists had increasingly turned to explaining social behaviors they attributed to contemporary humans as adaptations selected for during the Pleistocene. (Not all Socio-biologists took up this project, hence the qualification "Human" in "Human Sociobiology.") Famously, Sociobiologist Richard Dawkins broke ranks with his colleagues in *The Selfish Gene*, in which he argued that many aspects of human social behavior could not plausibly be construed as genetically based (Dawkins 1978, 189–201).

In explaining the rationale of seeking evolutionary explanations of human social behavior that invoke genes selected for during the Pleistocene, Wilson appealed to several assumptions and developments. One obvious and scientifically uncontroversial assumption is that human biology is the prod-uct of evolution. A second but controversial assumption is that evolution is also responsible for patterns of behavior that are characteristic of humans. More specifically, Human Sociobiology assumes that we have inherited specific genes that underlie specific social behaviors, and that these genes were "selected for" in the Pleistocene era because they conveyed an advantage in terms of survival or reproductive success. This is not to say, some human sociobiologists argue, that the genes we have inherited contribute to behaviors or features of psychology that are *now* adaptive. But qualifications such as these did little to assuage critics of the program who find biological, in this case, genetic, determinism to be deeply flawed. Moreover, critics charged that many claims made by human sociobiologists in the 1970s and 1980s did not embrace this qualification, but assumed that genes that were adaptive in the Pleistocene are still adaptive, an assumption we explore below.

Wilson also cited what was then known about causal relationships between specific genes or groups of genes, on the one hand, and behaviors exhibited by relatively "simple" organisms, on the other hand, as indicating that research into genes that underlie human behavior is warranted. He also noted the then relatively recent identification of specific genes contributing to or causing specific human diseases and conditions. Finally, Wilson argued that "universals" in human behavior spanning very different historical eras and cultures, constitute a *prima facie* case for the view that such human behaviors are biologically rather than culturally based.

In addition to proposing genetic bases for xenophobia and war, Wilson and other human sociobiologists devoted a good deal of attention to proposing adaptations of what they assumed or argued to be sex/gender differences in mating strategies and behavior. Here, they relied almost exclusively on PIT. They took it to provide explanations of why men are "promiscuous" and "undiscriminating" and women are "coy" and "choosy"; of rape (Barash 1979); and of societal practices and institutionalized norms that, in his book *On Human Nature*, Wilson described as a universal "Double Standard" that allows men to engage in a range of activities frowned upon or outright denied to women. And of "early man," Wilson wrote

> What we can conclude with some degree of confidence is that primitive men [*sic*] lived in small territorial groups, within which males were dominant over females ... Sexual selection would tend to be linked with hunting prowess, leadership, skill at tool making, and other visible attributes that contribute to the success of the family and *the male band*. (Wilson 1978a, 292; emphasis added)

Feminists have argued that Wilson's account of Pleistocene life, based as it was on earlier, ethnocentric and androcentric accounts of contemporary hunter-gatherer groups developed in the 1950s and 1960s in anthropology, including "Man, the hunter" theory of evolution, and on a selective group of primate species (most often baboons and ordinary chimpanzees), was deeply informed by gender stereotypes that in turn shaped observations of male and female behaviors. As noted in Chapter 2, feminists have argued that there is no basis for the assumption that our female ancestors were not contributing to the sustenance of their groups or that their activities were limited to those involving reproduction.

But Wilson's appeal to Parental Investment Theory in *On Human Nature* reveals it was an important, if not the most important, source of his assumptions about early human societies and about sex/gender differences among contemporary humans. Wilson cites several "consequences of gametic dimorphism." Here is one representative passage:

> The consequences ... ramify throughout the biology and psychology of human sex. The most important immediate result is that the female places a greater investment in each of her sex cells ... In contrast, a man releases 100 million sperm with each ejaculation.
>
> In most species, assertiveness is the most profitable male strategy ... It pays males to be aggressive, hasty, fickle, and undiscriminating. In theory, it is more profitable for females to hold back until they can identify males with the best genes. In species that rear young, it is also important for the females to select males who are more likely to stay with them after insemination.
>
> Human beings obey this biological principle faithfully. (Wilson 1978a, 129)

Sociobiologist David Barash was even more explicit about how we should understand the consequences of sex differences in "parental investment":

> Sperm are cheap. Eggs are expensive. Accordingly, females have a much greater stake in any reproductive act. Biologist G.C. Williams points out that in virtually all species males are selected to be aggressive – sexual advertisers – while females are selected to be choosier – comparison shoppers. Again, these behaviors follow directly from the biology of what it is to be male or female. (Barash 1979, 48)

Social scientists critical of Human Sociobiology, including but not limited to feminists, challenged the idea that the behaviors and forms of social organization on which Wilson and others focus, are trans-historical and cross-cultural "universals." Feminists also criticized the gender stereotypes that frequently informed research and hypotheses in the field; nor were they alone in doing so – philosopher Philip Kitcher criticized Human Sociobiology's reliance on what he called barroom gender stereotypes (Kitcher 1985). In addition, feminists argued that many of the behaviors and characteristics claimed to be universal and sex-differentiated fail to qualify as serious scientific "objects." What, they asked, are the behaviors supposedly captured by "female choosiness," which is alleged to be characteristic of females in all sexually reproducing species, including women? Although phrases like "male aggressiveness," "male promiscuity," and "female choosiness" may carry salience for those living in contemporary Western societies, what is their status as scientific concepts? Do they have clear identity criteria that span species, historical epochs, and cultures (e.g., Bleier 1984)? We will have reason to return to these questions and issues in forthcoming discussions of several research programs in biology.

The last criticism leveled by feminists that we consider is that, despite claims to the contrary (e.g., Wilson 1978b), some human sociobiologists have drawn normative conclusions (that is, claims about how men and women "should" behave) from PIT and genetic determinism. For example, Barash offered the following argument in a book written for the lay public.

> Because men maximize their fitness differently from women, it is perfectly good biology that business and profession taste sweeter to him, while home and child care taste sweeter to women.
>
> While it may be true that it's "not nice to fool Mother Nature", it can be done. Biology's whispers can be denied, but in most cases at a real cost . . . Although women who participate [in work outside the home] may be attracted by the promise of "liberation", they are in fact simply adopting a male strategy while denying their own . . . Cavalier female parenting is maladaptive for all mammals; for humans, it may be a socially instituted trap that is harmful to everyone concerned. (Barash 1979, 114–115)

This passage clearly includes normative claims. And they do not exhaust Barash's views about differences in male and female strategies for achieving reproductive success. He also suggested that rape may be a reproductive strategy. "Rape in humans," Barash notes,

> is by no means as simple [as the rape among mallard ducks I have observed], influenced as it is by an extremely complex overlay of cultural attitudes.

Nevertheless, mallard rape and bluebird adultery may have a degree of relevance to human behavior. Perhaps human rapists, in their own criminally misguided way, are doing the best they can to maximize their fitness. If so, they are not that different from the sexually excluded bachelor mallards. (ibid., 55)

Feminists and others have argued that "rape," outside of the human species, is an evaluatively thick metaphor – a topic to which we turn in the next section – rather than a literal empirical description of behavior; and that so-called evolutionary explanations of it in terms of reproductive success, minimize the violence and suffering associated with rape (e.g., Bleier 1984). Finally, many viewed this and other passages in books Barash wrote for a lay public to be irresponsible, given their purely speculative nature. There were, they pointed out, no genes identified as "coding for" rape.

That Human Sociobiology never succeeded in identifying the genes underlying specific social behaviors in humans is one reason that theory is of less interest now than it was in the 1970s and 1980s, when there were fiery debates between geneticists and other scientists who supported the program and geneticists and other scientists who opposed it. Evolutionary Psychology, which emerged in the early 1990s, has much in common with the Human Sociobiology, and we turn to it in a later chapter. Both emphasize what they claim to be sex-differentiated behaviors including mating strategies; rely on sexual selection and PIT; are committed to using contemporary hunter-gatherer groups to "reconstruct" features of the Pleistocene; and are committed to biological determinism. Human Sociobiology was strongly criticized for each of these commitments as well as for others.

Philosophical Issues

Biological/Genetic Determinism

Biological determinism, which includes but is not limited to genetic determinism, in the sense feminists and others criticize, is the theoretical and methodological assumption that many human behavioral and psychological traits are rooted in one or more aspects of biology. In later chapters, we will find that, in addition to research and theories considered in this chapter, there are other fields and research programs in the biological sciences that appeal to a version of biological or genetic determinism to explain sex/gender differences in aspects of human behavior and/or psychology. Biological determinism stands in contrast to two other (alternative) assumptions: 1) that in many species, especially in what Daniel Dennett calls "non-stupid species," many if not most behavioral and psychological traits are the results of interactions between biological and environmental factors (Dennett 1995); and 2) that biology has no role in determining many if not most behavioral and psychological traits, again at least in the case of non-stupid species. For present purposes, we can think of members of "non-stupid species" as those who learn from experience and adjust their behavioral and/or psychological traits, to the extent possible, based on experience. Some feminists also use the term "reductionism" to describe the approach referred to here as "biological determinism." They do so because the explanations in question propose that complex biological phenomena are caused by features of biology considered to be the most basic – for example, genes. But "reductionism" is used by philosophers to describe other commitments (for example, that "mental states" are, in the end, just "brain states"). So, we use "biological determinism" in describing the critiques feminists have offered of many different explanations that we consider in this and other chapters.

We have seen that Darwin, Parental Investment Theory, and human sociobiologists all assume that many sex- and sex/gender-differentiated traits are biologically determined. Darwin held that sex-differences in secondary sex characteristics are biological and came about because of greater selection pressures facing males. PIT attributes differences between males and females in psychology and behavior to sex differences in parental investment that are themselves the result of gametic dimorphism. Human sociobiologists focus on humans (although they often appeal to primate behavior to try to gain insights into the genetic underpinning of human behavior) and propose that genes determine human social behavior and forms of social organization. As earlier mentioned, in later chapters we will encounter other proposed biological explanations of sex and sex/gender differences in a variety of traits. We will also find that some research programs claim to provide, not biologically determinist explanations, but rather "interactionist" explanations of sex/gender differences – explanations in which some biological and some non-biological factors are appealed to. In such cases, it is reasonable to consider whether in fact non-biological factors are investigated in the research in question.

It is obvious why biological/genetic determinism is taken to be a significant issue by many feminists, and a key focus of their engagements with biology. To whatever degree an individual feminist views biology to be relevant to sex/gender, she or he is also committed to promoting and enhancing women's opportunities, aspirations, and achievements. That possibility requires that, contrary to Sigmund Freud, biology is not destiny. As important to those who engage in feminist science scholarship are issues concerning evidence and how it does or does not support hypotheses that significant sex/gender differences are based in biology. Too often, feminists maintain, there is insufficient evidence to warrant biological determinist theories, and lots of counter-evidence that challenges them. We will be considering such arguments in forthcoming chapters. Biologically determinist theories of sex/gender have a long history, as our discussion of Aristotle in Chapter 1 illustrates. And the nineteenth century was rife with alleged biological determinist explanations of women's "intellectual inferiority," including that their brains are smaller than men's – a difference in size that proponents of women's intellectual inferiority assumed was not a function of women having, on average, smaller bodies than do men. Some also argued that women have smaller parietal lobes, while others argued that they have larger parietal lobes; some maintained that women have smaller frontal lobes, while others argued that they have larger frontal lobes; and some argued that women's brains have fewer "quirls" than men's (Sayers 1982/1990). In Chapter 6, which focuses on sex/gender and medicine, we consider how physicians in the nineteenth century maintained that women's reproductive organs are the source of any illness or disease they suffered from, even if men are subject to the same illnesses and diseases. A partial list of twentieth century biological explanations of women's alleged "inferior abilities" to engage in math and science includes less hemispheric lateralization *and*, on another account, more hemispheric lateralization; a relative lack of exposure (compared to that of men) of circulating androgens during prenatal development; and neuroimaging studies claiming to find that brains are sex-differentiated. We study these twentieth century hypotheses in Chapter 7, which is devoted to feminist engagements with neurobiology.

Feminist historians and biologists have argued that there are relationships between the emergence of movements seeking women's equality and the development of biologically determinist theories that challenge feminists' arguments that women's abilities in a number of highly valued areas are not inferior to men's, and/or that challenge arguments that women are biologically suited only for the private and reproductive sphere. We consider representative arguments in forthcoming chapters.

Nor, feminists argue, are relationships between women's movements and arguments for biologically determinist theories of sex/gender a relic of the distant past. For example, feminist primatologist Susan Sperling notes that with some notable exceptions (including E.O. Wilson in *some* of his writings) advocates of Human Sociobiology's arguments about sex/gender differences often explicitly described such arguments as "disproving feminism":

> It is no coincidence that [Human] Sociobiology and the second wave of Western feminism were simultaneous occurrences. Early Sociobiologists clearly envisioned their new model as "disproving" feminism. The Sociobiologist Pierre Van den Berghe wrote: "Neither the National Organization for Women nor the Equal Rights Amendment will change the biological bedrock of asymmetrical parental investment." (Sperling, 1991)

From a similar perspective, philosopher Philip Kitcher described Human Sociobiology's approach to sex/gender differences this way:

> Sometimes the expression [of sexism] is tinged with regretful sympathy for ideas of social justice (Wilson), at other times with a zeal to *epater les feminists* (Van den Berghe). It is far from clear that Sociobiologists appreciate the political implications of the views they promulgate. These implications become clear when a *New York Times* series on equal rights for women concludes with a serious discussion of the limits that biology might set to women's aspirations. (Kitcher 1985, 6)

However, it is not at all clear that all research into sex/gender differences has been aimed at "disproving" feminism. Hence, it is appropriate to ask, and feminists do ask: if disproving feminism is not the goal, why have many scientists and philosophers viewed research into differences between men and women to be so important? After all, humans are far *less* sexually dimorphic than many other species. Some, including Mill, view the preoccupation with identifying ways in which women are inferior to men and/or with explaining sex differences in power and perceived "appropriate" roles and spheres for women on the basis of claimed biological differences, as deriving from the self-interest of men in justifying differences in the power and roles society affords to men and to women (see, e.g., Mill 1869). A thoroughgoing empiricist, Mill argued that, in order to seriously consider the claim that women are genuinely inferior to men in the various ways Darwin and others claimed, we need *to test this hypothesis* by providing women with equal opportunities in terms of education, property ownership, career choices, and so forth. Given, Mills argued, that no such social experiments had at the time been undertaken – experiments which, if women performed at a lower standard than their male counterparts, would have confirmed hypotheses about their inferiority – perhaps his male colleagues were actually worried that women wouldn't fail and a social revolution involving sex roles would follow.

An alternative and more nuanced view is that the fact that differences, real and assumed, between men and women are of such long-standing interest in biology reflects the historical, cultural, social, as well as scientific contexts within which biology is undertaken. Feminists' arguments can be understood to maintain that, whatever the reasons for historical and contemporary interest in sex/gender differences, the very fact that there has been and continues to be such interest shows that it is important to study and critically evaluate claims for the existence of such differences, and, to the extent there are such differences, to explore whether there are equally viable, non-biological, explanations of them. These issues figure in the discussions of many of the chapters to come.

Finally, we will consider how one area of research – studies of the effects of different amounts of circulating "sex hormones" (androgens and estrogens) during prenatal development on male and female fetuses – is taken to provide a useful "baseline" for understanding the effects of hormones more generally, rather than serve as the basis for biological determinist accounts of non-anatomical sex/gender differences. That is to say, there are reasons other than establishing a biological basis (in this case, differing amounts of hormones) for sex/gender differences in abilities, behavior, and/or temperament. But we will also find that some working in these fields do use the findings that result to argue for sex/gender differences in terms of one or more such traits.

In short, biological determinism is a complex topic. On the one hand, as we learn in Chapter 5, prenatal sex hormones do have a causal effect on sex- and sex/gender differences in the anatomical structures and processes related to reproduction. On the other hand, as we learn in Chapter 7, some biologists offer hypotheses that prenatal hormones cause additional sex-and sex/gender differences – hypotheses that, feminist biologists argue are not warranted by evidence.

Gender Stereotypes, Gendered Metaphors, and "Evaluatively Thick" Concepts

We have seen that feminists claim that "gender stereotypes" are assumed and/or argued for by Darwin, advocates of PIT, and human sociobiologists.

Here we introduce feminists' interest in and critiques of what they call "gendered metaphors" in biology. Metaphors abound in science, including biology. They are often taken to facilitate scientific theorizing and viewed as unproblematic *provided that* they are recognized *as metaphors* and not literal empirical descriptions of phenomena. Much ink has been spilled in efforts to define "metaphor." For our purposes, it is sufficient to understand metaphors as words or phrases that attribute a characteristic or quality that is applicable to one kind of object to a different kind of object to which it is *not* literally applicable. Consider the often-cited example that so and so "is drowning in paperwork." Of course, one cannot literally drown in paperwork; the phrase succeeds by likening the experience of being overwhelmed by paperwork with that of drowning.

We have discussed several metaphors in this chapter – including "parental investment," which brings economic theory to bear on the mating strategies of biological organisms; and descriptions of non-human animals as "coy" (typically, females) and as "promiscuous" and "undiscriminating" (typically, males); and "rape" in non-human species. Feminists argue that these linguistic expressions are metaphors, not literal empirical descriptions. And even if "coyness" and "promiscuity" are taken to be appropriate in the human case, they can only be metaphorically attributed to females and males of other species. And, even in the human case, we noted that feminists argue that these characteristics lack clear criteria and so are not respectable as scientific "objects."

Feminists further argue that gender stereotypes and gendered metaphors that are features of reasoning and hypotheses in research into sex/gender differences, also convey or include normative or evaluative content. Here the suggestion that some concepts are "evaluatively thick" is relevant. An evaluatively thick concept is a concept that contains *normative* (i.e., evaluative) content as well as empirical content (Anderson 2004). Some philosophers argue that there are such concepts, and that understanding their nature and role in our reasoning is important; these arguments were offered prior to feminist engagements with science and the second wave of the Women's

Movement. But the recognition of evaluatively thick concepts can help us understand feminists' concerns with the stereotypes and metaphors they find at work in areas of biological research, as feminist philosopher Elizabeth Anderson maintains (Anderson 2004). Arguably, the metaphors "coy females," "promiscuous males," "parental investment," "cheap sperm," "expensive eggs," and "rape in a non-human species," are such concepts. They carry normative content and implications, and thus are of interest and concern to feminists.

Has Feminism Changed Evolutionary Biology?

Most feminist biologists who address the question, "Has feminism changed evolutionary biology?" tend to say, "Yes, but we still have a long way to go" (Gowaty 2003). The most obvious change is in the number of women who have entered the field. More complicated is the question of whether their presence in substantial numbers has led to philosophical, theoretical, experimental, and/or practical changes. We consider this question here and continue our discussion of it in Chapter 4 where we consider arguments about the changes in primatology that some attribute to feminism.

Evolutionary biologist Patricia Adair Gowaty has surveyed developments in evolutionary biology that challenge PIT and she predicts additional research that already underway or envisioned will continue to provide evidence that undermines the theory. The seeds have been sown, Gowaty suggests, for important changes in how evolutionary biologists study and understand sex/gender. She foresees her colleagues undertaking, or in some cases *continuing*, detailed and species-specific studies of mating and parenting strategies, as well as of female and male behavior and temperament. These kinds of studies, Gowaty speculates, were not undertaken for many decades because the near universal acceptance of PIT, and for some, the "obviousness" of its hypotheses apparently led biologists to view such studies as unnecessary.

We have discussed feminist critiques of sexual selection and PIT in this chapter, and noted that some were based on field studies that failed to confirm the theory's predictions. Gowaty's survey indicates that in the decades since those critiques were offered, evolutionary biologists (many of whom do not self-identify as feminists) stepped up the pace of species-specific studies that continue to challenge the generalizations of PIT.

Gowaty is careful not to claim that feminism was the only, or even the primary, factor that brought about these specific studies, but she does believe it played an important role in bringing them about. She reports that disagreements remain concerning the relevance of feminism to evolutionary biology, some of which date to earlier "feminist discourse [that seemed] antagonistic to science." But she also notes that "tolerance" for feminist critiques and suggestions has become much more common in evolutionary biology. By "tolerance," she explains,

> I mean that ideas that come out of feminism are taken seriously [by many of my colleagues] as long as they are discussed in terms of testability, which is the hallmark of all ideas tolerated by scientists . . .
>
> Many of my colleagues pay attention to feminist-inspired scientific ideas – whether they come from women or men and even when investigators explicitly identify their sources from within late twentieth-century, [and] early twentieth-century Western feminism. (ibid., 902–903)

In addition, Gowaty takes the disagreements she mentions to include disagreements about whether "fundamental sexual natures" obviously exist, as some evolutionary biologists assume, and those who are skeptical of their existence (ibid., 902). She also is optimistic that this disagreement, which did not exist prior to the inclusion in evolutionary biology of large numbers of women and men who were open to feminist ideas, may stimulate further research on this topic. And in more general terms she argues that what she calls "feminist consciousness" on the part of both men and women has led to increased attention "to feminist issues . . . and is changing evolutionary biology" (ibid., 902).

Among the changes Gowaty cites are that it is now easier to publish "feminist-inspired hypotheses" and, more importantly, that such hypotheses are increasingly tested by men as well as women. And, she argues, bringing about "carefully designed, well-controlled, empirical tests," some of which are being undertaken and others that she is optimistic will be undertaken, is "the most efficient way to changing an entrenched scientific idea [in this case, Parental Investment Theory]" (ibid., 903).

Citing examples we have considered in this chapter, Gowaty notes other cases that challenge PIT's predictions about male sexual strategies and not just in those species for whom the predicted roles are reversed (i.e., males invest more in offspring than do females). As we will consider in the next chapter, female chimpanzees and langurs, and females of other primate species, solicit sex and are, in Gowaty's words, "anything-but-passive-coy-discrete" animals (ibid., 907). Indeed, some female primates have been observed to mate with multiple males during a single hour. This, Gowaty argues, is "hardly the behavior one would have predicted for a species with a long gestation similar to humans, a costly period of lactation (up to four years), and one of the longest periods of offspring dependence in nonhuman animals" (ibid., 907–908). So, too, male primates have been observed to be disinterested in sex, rather than "eager" and "promiscuous" as predicted by PIT. In each of the cases cited, observations were made over long periods of time of members of a single species, and as Gowaty notes, unlike earlier research in primatology, females were observed at least as much as males, let alone the "alpha males" on which earlier research often focused. As we will see in Chapter 4, feminist primatologists led the way in terms of studying female primates and their relationships to their offspring, arguing that what they described as "the female's perspective" needed to be studied and understood.

In addition to species-specific research that actually tests the hypotheses of PIT, Gowaty notes that alternatives to the theory itself have emerged. As early as 1985, evolutionary biologist William J. Sutherland proposed an alternative to Bateman's explanation (assumed by Trivers and PIT) of the differences in reproductive success he observed among male fruit flies, differences that were long assumed by evolutionary biologists to also be present in the majority of sexually reproducing species characterized by gametic dimorphism. Sutherland proposed a null model, according to which the differences Bateman observed could have come about by chance. Sutherland appealed to "handing time," which Gowaty describes as

> the time it took to copulate plus the time between the copulation with one partner and time to the onset of receptivity to re-mating, which might include parental care or regaining nutritional readiness to reproduce again, and found that male and female differences in such time could account for the sex differences in reproductive success Bateman observed. (Gowaty, 908)

Gowaty notes that although Sutherland's model was not generally pursued, it inspired Stephen P. Hubbell, a theoretical ecologist, and Leslie K. Johnson, an animal behaviorist (who in personal communication with Gowaty described themselves as feminists). They had previously studied so-called role-reversed beetles (the mating and parenting strategies of which are the opposite of what Bateman and PIT predict). Based on that research and Sutherland's suggestions, they created a mathematical model to explore the role of random environmental variation in leading to indiscriminate mating or, alternatively, to "choosy" mating. Their study took many variables into account and is understood by them, Gowaty, and some others to suggest that

> Mating strategies need not be due to sex differences fixed by ancient selection pressures. It showed when selection will be against choosy females and indiscriminate males, even in typical species with higher female than male parental investment. (ibid., 909)

Again, Gowaty notes that although Hubbell's and Johnson's hypothesis has not yet enjoyed general acceptance, it is only one of several alternatives to PIT being developed by evolutionary biologists and she provides details about several others. And Gowaty encourages evolutionary biologists to undertake empirical tests of these alternative models.

> Until experimentalists eliminate or control [alternatives such as] sperm limitation, latency to re-mating receptivity, and encounter rates ... conclusions about inevitable sex role differences determined by parental investment patterns are premature. (ibid., 912)

In addition, although the alternatives to PIT that have emerged (we have only discussed two of a number that Gowaty provides accounts of) have not enjoyed the degree of uptake in terms of testing that Gowaty had hoped, that they exist at all represents a change that Gowaty attributes to feminism and takes to provide reason to be optimistic about the chance that further changes will follow.

In her conclusion, Gowaty argues that "feminism changed sex role science in evolutionary biology in at least four ways." First, with the entrance of women into evolutionary biology "came their empathic responses to the lives of the nonhuman creatures they studied ... and facilitated the birth of the 'female perspectives' movement" (ibid., 916). Even those who, in Gowaty's words, "distance themselves from the politics of feminism," use the female perspective language. (She cites Parker and Burley 1997 as an example.)

Second, Gowaty maintains that as feminism itself spread, so too feminist consciousness grew among both male and female scientists. Such conscious-ness "nurtured the recognition of alternative hypotheses ... and to some scientists' insistence that the lives of females were worthy of study in their own right" (ibid., 916).

Third, Gowaty maintains that feminism allowed her and other evolutionary biologists "to frame testable hypotheses" that served as important alternatives to traditional approaches and hypotheses, many of which – including PIT – were not previously subjected to much by way of testing. One such alternative, offered by primatologist Barbara Smuts, is that male coercion and/or manipulation of females may also explain females' "choosier" behavior.

Lastly, Gowaty argues that feminism has led to better experimental design given that many more variables are now recognized as possibly contributing to mating and parenting strategies and because, at least in her own case, the awareness of her politics, causes her to make "my experiments better than they otherwise might be" (ibid., 916). More recently, Gowaty edited a collection of essays focusing on feminism and evolutionary biology (Gowaty 1997).

We next turn to feminist engagements with primatology, some of which many, including primatologists who do not self-identify as feminists, take to have led to important changes in observational methods and in accounts of the behavior and social organization of many primate species. The impact feminist engagements have had on primatology has led some to call the current field "a feminist science." As readers might guess, there are primatologists who, despite making use of the categories and methods feminist primatologists argued for in the 1980s, vigorously reject the notion that their research is "feminist" or in any way motivated by feminism. Rather, they contend, their research is just good science, and engaging in good science is what motivates them.

4   Primatology

Primates and Primatologists

We humans are primates. In the system of classifying living things initially devised by the Swedish naturalist Carl Linnaeus in the eighteenth century, "Primates" is an *order* of the *class* mammals, and includes such species as gorillas, chimpanzees, orangutans, langurs, monkeys, and humans, as well as extinct species. The discipline of primatology studies nonhuman primates in research undertaken both in laboratory and field research. But we will see that primatologists have often taken the data produced in such studies as providing insights into humans. In what follows, unless noted otherwise, 'primate' and 'primates' refer to nonhuman primates.

The name of the order is derived from the Latin *primus* meaning "prime," "most important," or "of the highest order." Species, both existing and extinct, have been added to the order as paleontologists, zoologists, and physical anthropologists, among others, have discovered them. The great apes, a sub-classification of primates that includes *Homo sapiens*, are more dependent for survival on learned behaviors, and in this and other ways are generally viewed as more intelligent than other primates, although we will see that primatologists who study other species, such as langurs and rhesus macaques, think otherwise.

Most primates live in social groups in which individual members regularly interact with one another. "Sociality," as primatologist Larissa Swedell refers to this characteristic of primate groups, brings with it benefits and liabilities (Swedell 2012, 1–2). The interactions that occur in primate groups (or, as they are sometimes called, 'troops') are

> Both affiliative (friendly) and agonistic (aggressive or submissive) . . .
> Individuals are gregarious; that is, they interact with one another

frequently, engage in a variety of types of social interaction, and typically form and maintain social bonds with other individuals. (ibid., 1–2; see, also, Dunbar 1988)

The "social bonds" Swedell describes include friendly bonds among a sub-group of a larger group of primates, which help the larger group in its agonistic interactions with other individuals or subgroups. And, despite its costs Swedell notes, "sociality is important to these animals ... [because] living in a group likely decreases one's risk of falling victim to predation" (ibid., 2).

Primate interactions, and how life within a primate group is "structured," are primary focuses of studies undertaken by primatologists. It is now recognized that patterns of behavior and social structures often vary by species, and sometimes by groups within a species (Fedigan 1986 and Strier 1994). However, during the 1960s and 1970s such differences were often not recognized. During that period, anthropologists studying savannah baboons took them to exhibit what they called "the primate pattern," a social structure and pattern of behaviors they assumed or argued to be characteristic of all primate species (e.g., Washburn and Lancaster 1968).

Although there are graduate programs in primatology, researchers in primatology are equally likely to have done their graduate work in biology, anthropology, animal behavioral science (ethology), psychology, zoology, or ecology. What demarcates the field from other areas of biology is its research focus – nonhuman primates, rather than training in a single academic discipline.

Primatologists undertake the study of primates, whether in the field or laboratory, for a variety of reasons. Some seek to learn about the behaviors and social dynamics of one or more specific primate species, or one or more groups of the same species. Others work to identify relationships between the behaviors and social dynamics of one or more primate species, on the one hand, and evolutionary selection pressures and processes, on the other hand. And among primatologists who focus on evolutionary issues in their study of one or more primate species, some seek to relate the behaviors and social structures of the primates they study to the evolution of human behaviors and social structures and/or to discover and explain aspects of "human nature."

Interest in studying primates to gain insights into human evolution and/ or "human nature" dates to the nineteenth century. Darwin, for example, took the morphological similarities between humans and African apes to suggest that humans evolved in Africa (Darwin 1871). Interest in relating primates' behavior and that of humans also characterized some research following Darwin's *The Descent of Man* (Tuttle 2014). Focusing on primates to learn about human behavior characterized some research in the early twentieth century, including the studies of great apes undertaken by Robert Yerkes and Clarence Ray Carpenter (Sperling 1991 and Tuttle 2014). But the emergence of comprehensive, empirical research devoted to relating the behavior, social dynamics, and evolution of primates to gain insights into human evolution and/or features of "human nature," did not emerge until after World War II, when American and European scientists began to undertake field studies in decolonized Africa and other Third World sites (Fedigan 2001 and Sperling 1991).

The rationale for looking to (non-human) primates for insights into human evolution, and/or into a possible "human nature," is straightforward. Several extant primate species – gorillas and two species of chimpanzee – are more closely related to us than is any other living species. It is widely accepted that we and they split from a common ancestor some six million years ago, a hypothesis supported by studies of the relevant genotypes. Individual humans differ from one another in only about 0.1% of their DNA. That is, any two humans have about 99.9% of their DNA in common. Comparisons between us and the two species of chimpanzees – so-called ordinary chimpanzees and bonobos – that focus on the same aspects of the genomes in question indicate that the differences between chimpanzee and human DNA is about 1.2%, and between gorillas and humans about 1.6% – about the same difference that obtains between gorillas and chimpanzees.

Our common ancestry, and genetic as well as morphological similarities with gorillas and chimpanzees, clearly warrant research into their evolutionary trajectory to learn about our own evolutionary history and its outcomes in terms of behavior as well as morphology. We might go further and assume, as some primatologists and Human Sociobiologists do, that these closely related primates can provide insights into a basic "human nature" that lies underneath the "overlay" of human culture – with culture understood as including verbal language, symbolic thought, the arts, social institutions, and the like. This assumes, of course, that there is such a thing as "human nature," a hypothesis that has critics as well as advocates.

For perhaps obvious reasons, the extent to which "biological" rather than "cultural" factors contribute to or cause human behavior and ways of organizing social life – the kinds of phenomena that those who assume there is "a human nature" have in mind – is far from settled. As we noted in our discussion of Human Sociobiology, one problem critics cited in terms of its efforts to identify universals in human social behavior and forms of social organization, is what certainly appear to be significant cultural and historical differences among human societies. Critics of efforts to use primates closely related to us to identify features of "human nature" point out that these primates are not our *ancestors*. Chimpanzees, gorillas, and humans have continued to evolve during the estimated 6 million years since we split from a common ancestor. Finally, critics of efforts to identify the sources of human behavior and forms of social organization through the study of primates point to the assumption of biological determinism that underlie them and offer arguments as to why the assumption is unwarranted. We will consider several such arguments.

On the other hand, it is difficult not to observe what appear to be significant similarities among the behaviors of some primate species and human behaviors. But substantive questions have been raised about some of the claims that general patterns of behavior and of social structure are common to many primate species and humans. One question involves the role and consequences of anthropomorphism. Some argue that many of the claims about similar behaviors and social structure would not seem plausible if the primatologists offering them had not interpreted the primate behavior, interactions, and "emotional" displays in question using terms we use to describe ourselves. We will see, for example, that feminists have been highly critical of the descriptions of primate behaviors and social dynamics involving gender that were common in the 1970s and early 1980s, on the grounds that they were anthropomorphic. (We will also see that the descriptions in question are far less common today and consider arguments as to why this is the case.)

But anthropomorphism in primatology is a complex issue. Not all primatologists view the attribution of human-like characteristics to primates to be inappropriate. Japanese primatologists (many of whom are men) view and "use [what they call] 'empathetic understanding' as a scientific tool" (Fedigan 2001, 62, citing Takasaki 2000). And several prominent women primatologists, whose work we will consider, argue that empathy is a tool that facilitates their understanding of female primate behaviors in the species they study.

However, these views are exceptions. The most common view among American and European primatologists is that anthropomorphism must be avoided. Feminist primatologist Linda Marie Fedigan's argument to this effect is representative.

> One of the strongest taboos in primate studies is to attribute human
> characteristics to animals – before we have studied them to determine how
> they do, in fact, behave and think – since this implies that all organisms
> behave and think like ourselves ... Primates are our closest relatives and share
> many characteristics with humans. But careful observations by scientists have
> convinced us that the gorilla does not beat his chest because he is angry; ...
> and the macaque does not grin because she is happy. Anthropomorphic
> assumptions about the meaning of primate signals must be avoided at all costs
> in that basic tool of all primate behavior research – the ethogram, which is a
> list of the behavioral units to be studied. (ibid., 56)

Fedigan notes that graduate students studying primatology invest a lot of time and effort learning how to use expressions "that are as value neutral as possible (e.g. 'open-mouth gape' instead of 'threat face')" (ibid., 56).

Still, Fedigan notes, avoiding anthropomorphism is difficult. As she makes the point, primatologists "are caught on the horns of a powerful language dilemma."

> We cannot fully invent a new language to describe our observations of
> animals, so we must borrow terms from the human domain that seem
> to best capture what we observe in the behavior of our animal subjects.
> (ibid., 55–56)

In forthcoming sections, we will have reason to consider the issue of anthropomorphism in primatology in more detail.

Primatology Following World War II

We previously noted that scientific interest in primates dates to the nineteenth century and was engaged in during the early twentieth century. But those who study its history date the emergence of primatology as a comprehensive, empirical, and quantitative science to the period following World War II. Although there is general agreement about the time frame, accounts of the research undertaken from the 1950s to 1970s differ in terms of which aspects of it were of most significance.

In "Reflections on a Century of Primatology" published in the late 1990s, W. Robert Dukelow, a former president of The American Society of Primatologists, emphasizes efforts by primatologists to protect primate species endangered by human activities, and to educate governments and the public about primates and the trends that threatened them (Dukelow 1999). He cites the founding of the "Regional Primate Research Centers Program" in 1960 as highly significant. It led to the funding of primate research centers in a number of locations in the United States, and Dukelow emphasizes the role of the centers in providing sanctuaries for displaced primates, fostering understandings of primate biology, and contributing to medical research (Dukelow 1999, 130). Dukelow also cites the creation of the International Primatological Society in 1964 as another significant development that contributed to primatology's emergence as a comprehensive science, as it facilitated international conferences "that would allow scientists and field workers to gather and exchange knowledge on nonhuman primates" (ibid., 130–131).

Dukelow's overview does not mention field studies during the 1960s and 1970s that sought to yield insights into human evolution and/or human nature. In contrast, feminists' accounts of the most important focuses and trends during the period do focus on this research priority. Primatologist Susan Sperling is representative in taking field studies of savannah baboons who lived in the Serengeti National Park undertaken by anthropologists working to identify "the primate pattern," as initiating what would become the dominant research priority of the 1960s and 1970s. The "pattern," which included behaviors and how social life was structured, these anthropologists argued, could be expected to characterize most primate species because it was rooted in selection pressures characterizing their evolution (Sperling 1991).

Sperling and others also cite field studies of so-called ordinary chimpanzees at the Gombe Reserve in Tanzania (now called Gombe Stream National Park) beginning in the 1960s as important. Data reported in studies of these chimps came to be well known and influential, in part due to the work and publications of Jane Goodall (e.g., Goodall 1965). And in some important respects, chimpanzee life appeared to be quite different from the accounts given of baboon life. But, for reasons we will explore, a research model emphasized in studies of savannah baboons of the Serengeti, and the accounts of "the primate pattern" they were taken to exhibit, is viewed by Sperling and others to be the most influential of the 1960s and 1970s, until Human Sociobiology became accepted (ibid., 8; also, Fedigan 1982, Haraway 1989, and Milan 2012).

Sperling notes that anthropologists of the period who argued that there is "a primate pattern," adopted a "structural-functionalist model" to explain the behavior and social dynamics of savannah baboons, and their reports and publications contributed to the adoption of the model in studies undertaken of other primate species. The model (which was not originally developed by anthropologists or other primatologists, but by social scientists such as Robert Merton for the study of human societies), assumes that "the structural form" of a social system functions to fulfill the needs of the individuals within the system, as well as to maintain itself. In each case, Merton argued, the important outcome is stability (e.g., Merton 1968).

Applying this model to the evolution of primates, anthropologists viewed the behaviors and "social structure" they observed in savannah baboons as "adaptations" (i.e., the product of natural or sexual selection) that functioned to promote the survival and reproductive success of individuals within the system, as well as to maintain the system itself, again by maintaining stability. Influential anthropologists, including Sheldon Washburn and Irven DeVore, took male dominance to be the most basic feature of savannah baboon social structure and related it to stability. Male dominance functioned, they argued, "to organize and control the troop in much the same way as political leadership functions in human cultures" (Sperling 1991, 1–2, citing DeVore and Washburn 1963, among others). Many of their contemporaries accepted the hypothesis that male dominance is central to primate social structure.

Sterling and others note that there was also strong interest among the anthropologists studying savannah baboons in extrapolating from their observed behaviors and social dynamics to human behavior and social dynamics. The factors they took to warrant such extrapolations included the belief that early hominids evolved in the African savannah under conditions similar to those in which the baboons lived, and the fact that humans are primates. Washburn's assumption of causal relationships between the evolution of specific species of nonhuman primates and that of humans, is clear in the following passage that appeared on the cover jacket of his book, *The Social Life of Early Man*.

> The social relationships that characterize man cannot have appeared for the first time in the modern human species ... Since man is a primate who developed from among the Old World simian stock, his social behavior must also have evolved from that of this mammalian group. Thus the investigation of man's behavior is dependent upon what we know of the behavior of monkeys and apes. (Washburn 1961)

Sperling and other feminists also note the emphasis anthropologists relating baboon behavior to human behavior placed on gendered behavior and gender relations – including similarities between male dominance in baboon groups and men's dominance over women, and similarities in what they described as "divisions in labor by sex" in both species. Sperling also discusses how anthropologists used savannah baboons to offer hypotheses about "the origins of 'the family'," an "entity" feminists argued that anthropologists described using gender stereotypes, and gendered and evaluatively thick metaphors (ibid., 7; see also Fedigan 1992 and 2001). A third and related emphasis in the studies of savannah baboons was what anthropologists described as "male aggression," a characteristic they also attributed to men and took, in the case of both species, to be evolutionary in origin (Sperling 1991). Interestingly, during the same period, the data reported about savannah baboons by primatologists such as Washburn and DeVore, was not substantiated either by studies of captive baboons or by studies of baboons in the wild that were undertaken at other sites (e.g., Rowell 1974).

As we have noted, during the 1960s and 1970s, extensive, long term studies were also undertaken of chimpanzees' behaviors and social organization in the Gombe Reserve in Tanzania, and they yielded a rich body of data. According to Sperling, the data were also "incorporated [by some primatologists] into structural-functionalist models for human evolution, which emphasized male dominance and aggression, and other gender-differentiated behavior" (ibid., 7). However, feminists, historians, and others point out that there were significant differences between the social dynamics involving gender reported by observers of savannah baboons and those reported by observers of chimpanzees of the Gombe Reserve.

The data provided about chimpanzees indicated that mother–infant relations are as central to group social structure as are male dominance, male dominance hierarchies, and male aggression. Researchers also reported that mother–offspring relationships were highly complex, and involved far more than gestation and lactation. They also described social relations among chimpanzees as being less hierarchical and more flexible than those reported to characterize relationships among savannah baboons (Sperling 1991, 10). Primatologist Donna Haraway summarizes what the reported data suggested about chimpanzees.

> Chimp social lives were stunningly flexible and complex, and chimpanzees seemed to require scientists' attention to individual actors as well as to social structures and biological and ecological parameters. Goodall's long-term studies of mother-infant relationships indicated a complexity and extent of maternal investment that would have been expanded even further in hominid evolution. (Haraway 1989, 337)

Nevertheless, many primatologists working in the period emphasized male dominance hierarchies, and male aggression, including in their accounts of chimpanzees and in reconstructions of human evolution. Indeed, data reported by those working at the Gombe Reserve did note that males were dominant over females, that there were male dominance hierarchies, and instances of male aggression. But, as we noted, their data suggested an equally important role for females and their interactions with infants, and that male dominance was not the most important feature of chimpanzee group life.

But in the end, the model of primate behavior based on savannah baboons as described by anthropologists did prevail during the period and was often extrapolated to humans. Although chimpanzees would become the more studied species in the late 1970s, male dominance and aggression reported in chimpanzee studies were emphasized rather than the other complex relationships, including the importance of female–infant relationships, reported by researchers studying them (Fedigan 1982 and Haraway 1989).

And so, a picture of primates, shaped by such perceptions, took hold. Primatologists Carol McGuinness and Karl Pribram, writing in the 1970s, reflected the widespread acceptance of "the primate pattern" initially proposed based on observations of savannah baboons, and reinforced by the emphasis that came to be placed on male dominance and aggression among chimpanzees, and of its applicability to humans.

In all primate societies, the division of labor by gender creates a highly stable social system, the dominant males controlling territorial boundaries and aggression, the females tending the young and forming alliances with other females. Human primates follow this same pattern so remarkably that it is not difficult to argue for biological bases for the type of social order that channels aggression to guard the territory which in turn maintains an equable [i.e., consistent and safe] environment for the young. (Quoted in Sperling 1991, 10)

As this statement makes clear, biology – rather than environmental, ecological, or social factors – was assumed to be the cause of "the primate pattern."

Philosophical Issues

We consider several philosophical issues in this chapter. We explore the issues relevant to a list of tools Londa Schiebinger recommends feminists make use of to analyze the relationships between gender and science that may obtain in specific research programs or sciences. We also consider feminists' general arguments for the claim that knowers are "situated" in relation to specific social contexts and in relation to their experiences within them. Although this is a general thesis to which many feminists subscribe, its relationship to arguments offered by primatologists Jeanne Altmann and Sarah Blaffer Hrdy considered in this chapter, makes its introduction here appropriate. We consider some of the ways in which feminists have sought to re-conceptualize objectivity in light of the thesis. Finally, we consider the question, "Has feminism changed primatology?" We begin with a discussion of Schiebinger's "tools of gender analysis."

Londa Schiebinger's Tools of Gender Analysis

As part of our discussion of Contextualism in Chapter 2, we introduced a model for analyzing scientific reasoning proposed by Helen E. Longino and Ruth Doell. Here we introduce aspects of a second model, developed by historian of science Londa Schiebinger, that also identifies tools feminists and others interested in relationships between gender and science can use to study them (Schiebinger 1999).

Like Longino and Doell, Schiebinger views the model she develops as capable, not only of illuminating unwarranted assumptions about gender in various aspects of research, but also as being able to bring about changes in the reasoning, social dynamics, and organizational structure of a research program when one or more of them involves gender. This can occur, she argues, by making the problems clear to scientists in the program. Her model has been influential in feminist science scholarship, and we will consider some of the ways that primatologist Linda Fedigan uses Schiebinger's tools to argue that, beginning in the late 1970s, feminist engagements with primatology led to significant changes in the field.

Some clarifications are in order. Schiebinger makes it clear that many of the tools she specifies as useful in analyzing how gender is related to some area of scientific research are not original. Tools of this sort were used by scientists in a number of fields and were seen as contributing to good research practices. Schiebinger's contribution is bringing the tools together and developing them in ways specific to gender. It is also important to note that Schiebinger maintains that, when the tools are used in analyses of the role of gender in a research program, they do not lead to "some special, esoteric, 'feminist science.'" Rather, they "incorporate a critical awareness of gender into the basic training of young scientists and the work-a-day world of science" (Schiebinger 1999, 186–190).

We will not be able to consider all the tools Schiebinger proposes (there are eight), or all the ways Fedigan and other feminists use them in their engagements with primatology. In the section devoted to feminists' critiques of and contributions to primatology, we organize the discussion using the tools of "research priorities," "representative sampling, "dangers of extrapolating research models," and "language use." Later, when we focus on changes in primatologists' approaches to gender that have occurred since the 1980s, we focus on the tool "the remaking of theoretical understandings." Although "the gender dynamics of a scientific discipline," which Schiebinger cites, is relevant to primatology, we include detailed discussion of it in Chapter 5 in our discussion of Developmental Biology. There are strong parallels in the gender dynamics of the two sciences. Here is a list of the tools on which we focus, or that are implicit, in discussions that take place in this chapter, and an example of the kinds of question each can be used to raise.

- Research priorities: this tool can be used, for example, to identify "what it is that we [in this case, scientists] want to know about."

- Representative sampling: this tool can be used to study if the subjects on which research focuses are sufficiently inclusive to represent the phenomenon or phenomena being studied.
- Dangers of extrapolating research models from one group to another: this tool can be used to ask whether it is appropriate to apply models of behavior and social dynamics developed in relation to one primate species to another – or, indeed, to all primate species,
- Language use: this tool can be used to ask if observations of animal behavior are couched in terms that are anthropomorphic, androcentric, metaphorical, and/or stereotypical.
- The remaking of theoretical understandings: this tool can be used to ask if feminist engagements with a field or science led to changes, in the sense of improvements, in the generalizations about sex/gender that once characterized research in it.

Feminist Critiques of and Contributions to Primatology

Research Priorities: Feminist Critiques

Schiebinger frames the issue of research priorities in terms of the question "*What is it that we want to know about?*" Feminists who critically engaged primatology beginning in the 1970s and 1980s argued that its researchers' priorities were androcentric, as they routinely focused on male dominance over females, male dominance hierarchies, male bonding, and male aggression (Fedigan 1992 provides an excellent overview of studies of primates that were so informed).

Feminists also criticized the priority given to gender relations and behavior based on a structuralist-functionalist model that emphasized male dominance, as well as on research priorities that assumed a Sociobiological model emphasizing Parental Investment Theory that came to dominate the field in the late 1970s. The result, feminists argued, were studies that did not sufficiently prioritize the study of factors that also impact gendered behavior and social dynamics. Such factors, feminists argued, include: differences in the size of primate "social groups" (these range from very small units to very large groups); the composition of primate groups (some are multi-male and multi-female, some are single male–multi-female, etc.); the role of learning in a primate group; and the impact of ecological factors.

Feminists also criticized the priority given to the kinds of mating systems in primate groups that seemed to fit with structuralist-functionalist models emphasizing male dominance, and with Sociobiological models emphasizing Parental Investment Theory. This priority was inappropriate, they argued, because primate species exhibit a variety of mating and parenting behaviors: some had been observed to be monogamous, some to be polygynous (these are groups in which individuals mate with more than one member of the opposite sex, including groups in which females routinely mate with more than one male). They also argued that prioritization of "dominance," "aggression," "divisions of labor by sex," and "female passivity" apparently led primatologists to make use of these categories even though they were not clearly defined or operationalized – that is, a pattern of behavior had not been specifically defined, nor were procedures for identifying and measuring it.

A second research priority, as we noted in the 1960s and 1970s studies of Savannah Baboons, was to gain insights into "men's aggression" by studying what was described as "aggression" exhibited by male primates. Some attribute this priority to an increasing concern with men's aggression as exhibited in World War II. But such interest, feminists argued, did not warrant the general lack of attention to the behavior of female primates.

A third priority was to learn about human evolution and possibly aspects of a basic "human nature" by studying nonhuman primates. As we have noted, such research focused on identifying "the functionality of gender relations," particularly male dominance over females; sexual divisions of labor; and the origins of "the family." Feminists argued that in most such efforts, androcentric and ethnocentric assumptions about human gender relations shaped the questions and hypotheses about, and the observations of, nonhuman primates, not the other way around. They also noted and criticized assumptions reflecting a commitment to biological determinism that informed such studies (e.g., Fedigan 1982; Zihlman 1978).

A useful way to characterize these critiques is to say that feminists sought to highlight the androcentrism, anthropomorphism, and biological determinism they took to characterize research in primatology in the 1960s and 1970s and, by so doing, change the research priorities of primatology.

Research Priorities: Feminist Contributions

Feminists argued that, in order to arrive at empirically adequate accounts of primate behavior and life, the research priorities of primatology would need to include attention to and knowledge about female primates: their perspectives; their contributions to the social dynamics of primate groups and not just their contributions to reproduction; and how mother–infant relationships, as well as male–infant relationships, might vary across primate species and be important features of a primate group's social dynamics (e.g., Altmann 1980; Fedigan 1982; Hrdy 1977 and 1986).

Other questions, feminists argued, also warranted study, questions that they began to bring to their own research. They include: Do the females of the species being studied form alliances with other females, and if so, why and how? Do such alliances have an impact on the social dynamics of the group, including on male behavior? Do females of a primate species compete with each other for mates and/or resources? If such competition does occur, is it informed by aggression, manipulation, or does it take some other form? Do some primate groups include female dominance hierarchies? Are there primate species in which females are dominant over males? Do females of a primate group ever lead, rather than merely follow, their male counterparts, when the group moves to forage or hunt? Are female primates (typically or always) "passive resources" for males in terms of mating, or do some female primates actively solicit sex? Do they use such solicitations to manipulate males to promote their own reproductive success? Do the females of some primate species, like males of some primate species, choose to mate with more than one partner (e.g., Haraway 1989; Hrdy 1986)?

And, feminists asked, shouldn't providing operational definitions, or information about relevant biochemical or physiological bases, for terms such as "active" and "passive," and "dominance" and "aggression," be a priority? Fedigan and other feminists argued that terms such as these had not been so defined or clarified; rather, they appeared to be "evaluative metaphors" serving (whether consciously or not) "to dramatize a particular interpretation" of primate behavior (Fedigan 1992, 23–24).

In more general terms, feminists asked what might we learn about the social dynamics of a primate species or group if we sought to find out "how things look from a female's perspective," rather than emphasizing those of, or at least those attributed to, males? And how might what we learn change our understandings of the behaviors and social dynamics we want to understand? (e.g., Altmann 1980; Hrdy 1986; Lancaster 1973).

Finally, feminists argued that if we assume that the study of nonhuman primates can provide insights into the selection pressures on ancestral humans and their effects on human evolution and contemporary human behavior, including those involving sex/gender, what issues of concern *to women*, including women who are primatologists, warrant attention? Barbara Smuts specifically raises such issues (e.g., Smuts 1992).

I have listed the questions cited above under "feminist contributions" to research priorities. This is how feminists view them, as they take them to have led to significant changes in primatologists' approaches to gender beginning in the 1980s, changes which resulted in more empirically adequate accounts of the social dynamics characteristic of several primate species.

Jeanne Altmann was one of the primatologists who chose "to study females, and to take 'the female point of view' as part of [a] feminist approach to primatology" (Fedigan 2001, 49). Altmann worked for several years in a long-term study of the Amboseli savannah baboons. Over time, those engaged in this large and intensive study had collected systematic data about social behavior, ecology, and demography; and had constructed mathematical models to deal with each and with relationships between them. Data collected before Altmann joined the research team were available to her in the form of a computerized data set housed at the University of Chicago (Altmann 1980).

According to Altmann, her study of female primates – in particular, the relationships between female baboons and their offspring – was something she initially wanted to avoid. She was worried that the research emphasis would be perceived as reflecting a non-objective "empathy" with her research subjects (interview with Donna Haraway detailed in Haraway 1989). Altmann also worried that studies of females would be viewed as unimportant given the then current emphasis in the field on male primates. In addition, she remarks that, in the 1970s, for a scientist to acknowledge that her research was in part related to her feminism, would be seen as deeply problematic (and, of course, some scientists and philosophers of science still hold this view). Feminism was "political" and value laden, the antithesis of what many thought science is and should be.

Yet, Altmann notes, eventually her own complex self-identity – "as a scientist, feminist, and mother" – led her to focus her research on female baboons. "Increasingly," Altmann reported in an interview, "it was screaming at me. These are the most interesting individuals; [interactions between baboon mothers and offspring] have the most evolutionary impact; this is where the ecological pressures are" (interview detailed in Haraway

1989, 312). In her field reports, Altmann described baboon mothers as routinely engaging in what she described as "juggling" between competing priorities and doing several things at once. She drew parallels between such competing priorities and those facing "dual-career" women, women who like herself are mothers and have careers (Altmann 1980). Altmann also notes that her studies of adult female baboons did not include the category "child care" because she could not find a way to separate females' activities in ways that would warrant that specialized category. As Haraway reports of her interview with Altmann, "Altmann could see no clear separation . . . between nursing, eating, carrying, disciplining, socializing with other adults, and paying attention to opportunities and dangers" (Haraway 1989, 313).

Using the metaphor of "budgeting" – and being quite specific that this *was* a metaphor – Altmann took her observations as indicating that female baboons' budgeting, and their interactions with juveniles as well as female and male members of their troop, were key factors in the group's social dynamics (Altmann 1980). Of budgeting, she told Haraway, "You could call it budgeting or something else . . . The issue is hierarchy of demands and the immediate consequences of these demands. What is and is not flexible in one's life is to me terribly fascinating, terribly important biologically, and also important for my experience as a human being, a woman, a mother" (interview in Haraway 1989, 314).

Altmann also reported complex interactions between adult males and infants among the baboons, observing that females developed special relationships with some males who then carried her infants in times of danger and protected them from other baboons (Altmann 1980). As primatologist Sarah Blaffer Hrdy describes the findings of Altmann's studies of these relationships in savannah baboons, "Infants, then, are often the focal-point of elaborate male-female-infant relationships, relationships that are often initiated by the females themselves" (Hrdy 1986, 20).

Further breaking with the emphasis on male-centered analyses of baboon social dynamics, Altmann proposed that the "high drama" characterizing encounters and relationships between male baboons, once the primary focus of anthropologists, is not nearly as important to the social dynamics of baboon groups as the "micro-practices" of females (Altmann 1980).

Feminist primatologist Sara Blaffer Hrdy has written articles and books based on her studies of hanuman langurs (her primary research interest), rhesus monkeys, and other primates (e.g., Hrdy 1977, 1986, and 2009). In each case, her studies focused on female primates. As with Altmann's observations of female baboons, the results of Hrdy's studies of female behavior stand in stark contrast to those once thought to be representative of the "primate pattern" emphasizing male dominance and describing females as sexually coy and subordinate. What sharply distinguishes Hrdy's methodological and theoretical approach compared with that of other feminist primatologists, is that she assumes and appeals to Sociobiology, including Parental Investment theory – theoretical models we have seen that other feminists are highly critical of. But Hrdy's accounts of female primate behavior are fundamentally at odds with those offered by primat-ologists and evolutionary biologists who accepted Bateman's, Trivers', and Wilson's view of how greater parental investment would affect females' behavior.

Hrdy maintains that it is precisely *because* of their greater investment in offspring that the female primates she has observed are far from coy, far from passive, and far from subordinate to their male conspecifics. In an article published in 1986, for example, Hrdy describes the results of her studies of female species as indicating

> That a polyandrous component [the term is used here to refer to a female with more than one established mate] is at the core of the breeding systems of most troop-dwelling primates: females mate with many males, each of whom may contribute a little bit toward the survival of offspring. Barbary macaques produce the most extreme example (Taub 1980), but the very well-studied savannah baboons also yield a similar, if more moderate pattern. (Hrdy 1986, 125)

Hrdy cites several ways in which female langurs manipulate males. Female langurs (and this is also the case in other primate species, Hrdy notes) can conceal estrus or bring about cyclical estrus to promote their reproduct-ive success. Contrary to hypotheses about how so-called "alpha males" have the most sexual access to females, Hrdy notes that females often solicit males other than dominant males. They do so, she observes, in three kinds of circumstance: when females leave their natal group to temporarily join an all-male band and mate with its members; when a female as she puts the point, "for reasons unknown to anyone, simply takes a shine to the resident male of a neighboring troop"; and when males from nomadic all-male bands join their troop temporarily (ibid., 126).

Again challenging what many if not most Sociobiologists took to be clear implications of Parental Investment Theory, Hrdy reports that in langurs it is *female-on-female* competition that is the most prevalent and consequential agonistic feature of langur group dynamics. "The manipulation hypothesis" Hrdy makes use of to explain aspects of female langur behavior, and that of females in some other primate species, is just one of six hypotheses that emerged in the 1980s to explain female behavior in primate species that directly contradicted the predictions of Parental Investment Theory – as the theory was understood by its advocates in the 1960s and 1970s.

As Hrdy describes the manipulation hypothesis, "females were more political, males more nurturing (or at least not neutral), than some earlier versions of sexual selection theory would lead us to suppose" (ibid., 129). And she argues that this hypothesis leads to previously unasked questions.

> Assuming that primate males do indeed remember the identity of past consorts and that they respond differently to the offspring of familiar and unfamiliar females, females would derive obvious benefits from mating with more than one male. A researcher with this model in mind has quite different expectations about female behavior than one expecting females to save themselves in order to mate with the best available male. The resulting research questions will be very different. (ibid., 134)

We have noted that some feminist primatologists criticize Hrdy's adoption of and appeals to Sociobiology and Parental Investment Theory, despite the very different picture of the power and influence of female primates yielded by her approach. Susan Sperling's critique is representative in stressing that Sociobiological approaches, like structural-functional approaches, seek to identify "ultimate" causal factors (they assume, in other words, biological determinism), and fail to sufficiently take "proximate causes," such as relationships between behavior, ecology, and social structure, into account (ibid., 1–4). Sperling and others also reject the "re-working" of what they take to be problematic models, such as Sociobiology, to re-conceptualize female primates.

> [Feminist sociobiologists] use the old narrative structures [and thus] tell us little about the development of complex behaviors and their context-dependent expressions ... [Is] it advantageous merely to change one narrative element, as feminist sociobiology has done, so that the category "female," like "male," is constructed as active, dominant, and looking out for genetic

advantage? I think not, and I want to argue instead for a deconstruction of all functionalist models, including sociobiological ones, of sex-linked primate behavior. (Sperling 1991, 4)

The issues involved in the question of whether scientists do or should seek "ultimate causes" of phenomena or (in the exclusive sense of "or") should study "proximate causes" of phenomena, is complex. In physics, there is a long history of searching for ultimate causes. There are also biologists who view their research as devoted to identifying ultimate causes of the phenomena they study. Obviously, Human Sociobiologists such as E.O. Wilson, who propose that genes selected for in the Pleistocene determine human social behavior, seek to identify ultimate causes. In contrast, the biologist Ernst Mayr argued that seeking "ultimate causes" to explain the traits of complex biological organisms is inappropriate given the complexity of such organisms, including the relationships between the parts and processes that characterize them, as well as causal effects of experience and environments on biological organisms (Mayr 1989). But a third perspective is also possible and argued for, namely, that in biology, scientists should seek both the ultimate and proximate causes of the phenomena they want to understand (Michael Ruse, private correspondence). We return to these issues in forthcoming chapters.

The continuing and expanding study of primate species in which feminist and other primatologists are engaged, including the changes Fedigan notes, may result in a field that remains heterogeneous in terms of its models and research questions, or primatologists may come to adopt some new model. I would only add that from my own reading of Hrdy's work, which focuses on several primate species and human females, it seems clear that she neither fails to take context into account nor "essentializes" females based on some "ultimate" cause of their behaviors. Females' greater investment in offspring leads to a variety of strategies depending on factors such as ecology, whether females are able to manipulate males in ways earlier cited, group size, and male parental investment. The variations Hrdy recognizes and attributes to female primates and women are suggested in the title of one of the sections of the article on which we have focused; the subtitle is "A Female is not a Female is not a Female" (Hrdy 1986).

We have focused on the research of only two of many feminist primatologists. The collection, *Female Primates: Studies by Women Primatologists* edited by

Meredith Small provides details about the studies undertaken by women primatologists, including feminists (Small 1984).

Representative Sampling: Feminist Critiques

The criticism of the relative lack of attention to female primates that we considered under the topic "research priorities" also involves the issue of "representative sampling." Feminists argue that for many years, primatologists, including those who sought insights into human evolution and behavior, appear to have been selective in terms of which species they studied.

As Fedigan notes, there are over two hundred primate species. (Indeed, The International Union for Conservation of Nature recognizes a total of 612 species and subspecies of living primates, and expects more to be discovered.) Why, feminists asked, were savannah baboons and so-called ordinary chimpanzees the species most studied to understand the alleged "primate pattern" and to gain insights into human evolution and sex/gender relations? Was it the male dominance, male dominance hierarchies, and male aggression attributed to these species at least part of the reason? Why weren't primate species characterized by far less sexual dimorphism in terms of size and strength (for example, there is far less dimorphism in langurs than in savannah baboons) studied to the extent that savannah baboons and (ordinary) chimpanzees were?

Similarly, why weren't primate species in which females mated with more than one male, or those in which male dominance hierarchies were less important features of social dynamics, also studied in detail? Were the primates selected for study for insights into humans chosen because they appeared to exhibit behaviors researchers assumed to be characteristic of humans? Why were the gender relations and social dynamics attributed to savannah baboons in the 1960s and 1970s generalized to all primate species, despite the evidence provided by study of chimpanzees at the Gombe Reserve that showed that those chimpanzees, although male dominant, did not fit "the primate model" constructed using savannah baboons in other important respects? Why wasn't there "uptake" of the data gathered at the reserve that indicated that female–infant relationships and female alliances are central to social dynamics?

Feminists argued that there were other examples of non-representative sampling. One of the more striking examples involves bonobos, a species of

Figure 4.1 Bonobo female grooming adult son, *Pan paniscus*, Congo, DRC, Democratic Republic of the Congo image 148307626.
Frans Lanting, Getty Images.

chimpanzee as closely related to humans as ordinary chimps. They had not been extensively studied for insights into human evolution, human nature, or the origins of human gendered behaviors, mating and parenting strategies, and gender relations. This is important, feminists argue, precisely because bonobos exhibit behaviors that stand in sharp contrast to those of the ordinary chimpanzee researchers have emphasized, and their groups are socially organized in strikingly different ways (Figure 4.1).

Primatologist Frans de Waal, who has studied bonobos and written about them for decades, also finds the relative lack of attention to them both worthy of note and in need of explanation. In the first chapter of *Bonobo: The Forgotten Ape*, de Waal describes the differences between the models of "primate behavior" based on savannah baboons, on the one hand, and human behavior and social dynamics, on the other hand, that eventually led many primatologists to turn their attention to ordinary chimpanzees to gain insights into humans.

Not too long ago, a much more distant relative [than ordinary chimps or bonobos], the savanna baboon, was regarded as the best living model of ancestral human behavior ... The baboon model was largely abandoned, however, when it became clear that a number of fundamental human characteristics are absent or only minimally developed in them, yet present in chimpanzees. (de Waal 1997, 3)

Among the characteristics common to humans and ordinary chimpanzees, de Waal cites food sharing, politics involving power, and tool use. He also describes the increased attention to (or what he called "selection of") the ordinary chimpanzee as the touchstone of human evolution as an important improvement over the savannah baboon.

But de Waal also notes that gender relations, and aspects of social dynamics involving gender and dominance, as they were reported among ordinary chimpanzees "did not need to be adjusted: male superiority remained the 'natural' state of affairs. In ordinary chimpanzees, as in baboons, males are conspicuously dominant over females" (ibid., 4). This is not the case, he notes, when we turn to bonobos.

> Enter the bonobo, which is best characterized as a female-centered, egalitarian primate species that substitutes sex for aggression. It is impossible to understand the social life of this ape without attention to its sex life: the two are inseparable. Whereas in most other species, sexual behavior is a fairly distinct category, in the bonobo it has become an integral part of social relationships, and not just between males and females. Bonobos engage in sex in virtually every partner combination: male-male, male-female, female-female, male-juvenile, female-juvenile, and so on. The frequency of sexual contact is also higher than among most other [nonhuman] primates. (ibid., 4–5)

Indeed, de Waal argues, studying the sexual activities, behaviors, and gender relations of bonobos would likely result in accounts of primate sexual relations that more accurately represent the *varieties* of forms they take, a point paralleling that of feminists who observed females soliciting mates in some primate species and often mating with more than one.

de Waal addresses what he sees as a possible explanation for the relative lack of attention to bonobos.

> The bonobo, with its varied, almost imaginative eroticism, may help us see sexual relations in a broader context ... Because the role of sex in society is such a loaded and controversial issue, scientists have tended to downplay this side of bonobo behavior ... If the apes themselves are so relaxed about it, it seems inappropriate for us to give in to typically human obsessions. (ibid., 5)

Moreover, de Waal notes, "there is a lot more to bonobo natural history than sex." The ways in which females and males behave and interact, including the female-centered and egalitarian ways of bonobo life, the ways in which juveniles are parented, and the relative lack of male aggression (which if threatened is promptly defused by females as well as other males), render "the entire social organization of the species ... fascinating" (ibid., 5).

Feminists have argued that attention to the ways in which female bonobos exert control over aspects of group life, and the prevalence of cooperative, rather than aggressive, behavior among bonobos, might have led to recognition of significant variation in the social dynamics that characterize primate species. Such recognition, in turn, might have led to studies into variations in human sexual behavior and forms of social organization, and to explorations of how social, rather than evolutionary and biological factors, may have constrained both (Fehr 2011).

Representative Sampling: Feminist Contributions

In terms of feminist contributions to sampling methods in primatology, work undertaken by Jeanne Altmann was particularly important and influential in the field. In an article published in 1974, Altmann undertook an extensive analysis comparing and contrasting sampling methods that were common in primatology: codifying them; identifying what, in terms of each, was required if it was to result in representative data; and identifying the limitations of each. Fedigan notes of Altmann's analysis that "it standardized sampling practices in ethology and ... it is still referenced in most methods sections of etholological publications" (Fedigan 2001, 51).

Altmann argued that one particularly problematic method used in the field was "ad lib sampling," which involved a primatologist recording whatever he or she happened to notice and find interesting. "Ad lib," opportunistic sampling, Altmann argued, cannot yield accurate comparisons of the rates individual animals engage in a specific behavior, although it was often taken to yield such results. Altmann also emphasized the need for longitudinal studies, and the need for sampling over representative periods of time, of the behavior of less-dominant males, females, and juveniles (Altmann 1974). Although Altmann did not describe her analysis as motivated by an interest in encouraging observations of female behaviors and their effects on social dynamics, that analysis did result in significant changes in the amount and quality of the observations devoted to females. Fedigan's description of the effects that Altmann's analysis had on approaches to gender in primatology is representative of analyses of the impact of Altmann's work offered by feminists.

> What this meant for primatologists is that we stopped watching only the larger, more swashbuckling males and started to also sample for representative periods of time the less prepossessing subordinate males, the females, and the immature individuals of the group ... What Altmann did ... was to convince scientists, through the use of their own methodological tools, to raise their standards of evidence, a task for which she was well qualified by her background training as a mathematician. (Fedigan 2001, 51; see, also, Haraway 1989)

Some years later, as part of her interview with Haraway, Altmann did relate her research priorities in terms of sampling to her feminism.

More Philosophical Issues

In this section, we discuss two related epistemological topics: feminists' arguments that knowers, including scientists, are "situated," and the implications of this for how objectivity can and/or should be understood.

Knowers as "Situated": Does "Who" Is Theorizing Matter to the Content of Science?

Three arguments became common in feminist science scholarship beginning in the 1980s. One maintained that the widespread presence of androcentrism in the biological and social sciences well into the 1980s (and still present in some fields), was due to the prevalence, before the lifting of formal and informal barriers to women, of men in the relevant sciences. In a second and related line of argument, feminists cited social beliefs and values concerning sex/gender common in the larger social and cultural contexts in which science is undertaken, as contributing to androcentric research questions, observations, and hypotheses.

More controversially in terms of traditional views of science, in a third line of argument feminists attributed the influx of women (and men) whose political beliefs and values were in part shaped by the Women's Movement, as enabling the recognition of androcentrism and of the alternative questions, methods, and hypotheses that feminist scientists began to propose.

In primatology, however, it wasn't the case that women had been vastly under-represented in the field prior to the late 1970s. To the contrary, as Fedigan puts the point, women were on the ground floor of this field by the mid-twentieth century (Fedigan 2001, 55). So, the more appropriate question in this case is whether feminism made a difference to the research priorities and interests, research questions and focuses, of some primatologists who entered the discipline in the 1970s?

All three arguments, and the question just identified, are controversial in some quarters. It was long accepted by most scientists and philosophers that the social identity of scientists – their gender, race, sexuality, culture, politics, and the like – is irrelevant to their science: to the questions they pursue, categories they assume, observations they make, methods they employ, and hypotheses they propose. Many believed, and continue to believe, that scientific methods, shared epistemic values taken to be "constitutive" of science, peer-review mechanisms, and other features of the way science is structured and organized, filter out any "bias" or unwarranted assumption an individual scientist or group might unwittingly bring to bear on their engagements with nature. (This assumption was reflected in the distinctions many philosophers drew between the contexts of discovery and justification, a distinction that as we have discussed, feminists challenge based on their analyses of scientific reasoning in areas of biological research.) Such intersubjectivity, together with the longstanding and related assumption that science is "value-free" (apart from the role of theoretical or cognitive values), seemed to rule out the possibility that who is theorizing *could* matter – at least that it could matter in *good* science.

But feminist scientists and science scholars view much of the research we have considered in this and earlier chapters on which feminists focus, to indicate otherwise. Good science, feminists take their research to indicate, can reflect and often has reflected the specific historical, social, and scientific contexts within which it is undertaken. (This reflects the Contextualist approach many feminists bring to their study of science that we discussed in some detail in Chapter 2). But feminists also argue that *within* the social contexts of Western societies, divisions in experience along the lines of gender, race, socioeconomic status, sexuality, and other socially salient categories, can also impact experiences and background assumptions relevant to scientific research, reasoning, and findings. The result, many feminists argue, is that knowers including scientists, *qua* scientists, are *situated* not only in relation to general scientific and social contexts, but also in terms of more specific social situations, such as gender. These factors feminists argue, can have consequences for their perspectives, background assumptions, and even for what they are able to know and/or what they claim to know.

In its general outlines, the thesis that knowers are "situated" is a central theme in feminist science scholarship, but both it and the consequences taken to follow from it, remain matters of ongoing analysis. Its strongest version, advocated by Feminist Standpoint Theorists such as Sandra Harding, maintains that differences in power along the axes of gender, race, culture, and other categories, provide opportunities for those with less power to recognize problems in the arguments taken to support their different roles and/or inferiority. As the scientists and philosophers whose views about "situatedness" do not, for the most part, appeal to Standpoint Theory, we do not emphasize it; but readers should know that it has been influential in feminist theorizing (e.g., Harding 1986 and 1991). We begin with arguments that challenge the claim that scientists are "situated" in ways that are consequential for science, and consider feminist responses to them.

To critics of feminists' arguments linking aspects of the broader social contexts within which science is undertaken to aspects of scientific research, neither the recognition of androcentrism nor of the problems to which it led, had or has anything to do with *feminism* and/or with "who" scientists are in terms of their social identities (e.g., Haack 1993).

Three arguments for this conclusion are common. In the view of feminists, the least plausible is that it was simply a *coincidence* that androcentrism was recognized when women entered science in larger numbers and/or female and male scientists brought a new sensitivity to issues involving sex/gender to bear on research in their fields. A second argument is that the recognition of androcentrism is an example of science's tendency to "self-correct" and, again, bears no relationship to feminism.

Sarah Blaffer Hrdy challenges the "coincidence" and the "self-correcting" arguments by noting the factors that led to some important changes in primatology.

I seriously question whether it could have been just chance or historical sequence that caused a small group of primatologists in the 1960s, who happened to be mostly male, to focus on male-male competition and on the numbers of matings a male achieved, while a subsequent group of researchers, including many women (beginning in the 1970s) started to shift the focus to female behaviors having long-term consequences for the fates of infants. (Hrdy 1986, 136)

In highlighting two different groups of researchers, Hrdy can be understood to effectively undermine both the "coincidence" argument and the "self-correcting" argument. The anthropologists focusing on males in the 1960s and 1970s did *not* "self-correct" their science. They were not responsible for the shift to studying female primates, to avoiding gender stereotypes in their accounts of primate behavior and social dynamics, and so forth. We have also reviewed arguments offered by other primatologists that feminism made a difference to the research in which they engaged, including arguments by Altmann and Gowaty. And Hrdy herself offers such arguments about her own research. These arguments are representative of those that feminist biologists in other fields have offered about their research priorities (e.g., Gilbert 1988, Bleier 1984, and Fausto-Sterling 1985).

A third line of argument maintained that the problems feminists have identified only involve cases of "bad science" and so are not relevant to "science as usual." But, feminists argue, androcentrism and other problems they cite do occur in scientific reasoning and hypotheses that meet criteria taken to characterize good science. Because of this, feminists argue, the science they criticize cannot be written off as "bad science."

If Scientists Are "Situated," What Are the Implications for Scientific Objectivity?

Recall that it was long assumed that the intersubjective nature of scientific theorizing – a function of the universality of the logic or reasoning characteristic of science, of the firm foundations that observations provide for scientific claims, of science's "value freedom," and of the importance attributed to peer-review and replicability of results – was the reason that science is objective. But the "situatedness" feminists attribute to scientists makes it possible, if not likely, that scientists bring background assumptions to their practice that they do not recognize and/or have not examined – assumptions that can help shape their research questions, observations, and hypotheses.

Does it follow that the very idea of scientific objectivity is a non-starter? Most feminists do not think so. Feminists agree with Hempel that background assumptions are an unavoidable aspect of scientific theorizing, and like many philosophers and historians of science, they reject the possibility of "value-free science." Indeed, most feminists recognize the role of values in their own theorizing within and about science. Among the concerns feminists emphasize in their discussions of objectivity is the need to address the role of unexamined and/or unwarranted background assumptions in mediating the relationship between available evidence and a hypothesis, and the importance of precluding what Helen E. Longino refers to as "individual subjective preferences" from determining the outcomes of scientific research. These and other issues relevant to objectivity we have considered and will continue to consider can be understood as reflecting a belief that evidence must be given its due, while also recognizing that because of their situatedness in specific scientific and social communities, scientists may not be aware of some relevant evidence or may not accept what some cite as evidence relevant to their theorizing.

In the 1980s, some feminists argued that objectivity requires that scientists critically scrutinize the empirical warrant of the background assumptions they bring to their practice, including the values these assumptions might presuppose. One obvious question, of course, is whether if one is a member of a relatively monolithic scientific community in terms of socially salient categories such as gender, it is likely that one's background assumption and/or social values are shared in the community and are thus unlikely to be recognized as in need of scrutiny.

This view of objectivity emphasizes scientists *qua* individuals, and some feminists point out that the recognition of unquestioned background assumptions, let alone of the need to critically examine them, would likely only occur in scientific communities sufficiently diverse in terms of the experiences and background assumptions of its members (Harding 1991; Longino 1990). But even this is only possible, many feminists argue, if science communities give sufficient recognition to views that differ from those dominant in the community. Many point out that in many science communities, "cognitive authority" is not shared equally.

In 1990, Helen E. Longino proposed an account of what is needed to foster the possibility that the knowledge a science community generates is objective to some degree or other: that, to whatever degree possible, "individual subjective preferences" do not determine the knowledge generated. What distinguishes Longino's account from those so far considered is that she emphasizes that the organizational structure of scientific communities, rather than efforts on the part of individual scientists, could be designed in ways that would contribute to the objectivity of its research results. "Criticism from alternative points of view and the subjection of hypotheses and evidential reasoning to critical scrutiny," Longino argued, "limits the intrusion of individual subjective preferences into scientific knowledge" (Longino 1990, 76). The criteria she identifies as enabling intersubjective criticism include 1) public venues for critical engagements, including "such standard and public forums as journals, conferences, and so forth"; 2) shared public standards, "both substantive principles and epistemic, as well as social values," that members of the community come to accept; 3) uptake by others of the criticisms offered of data and/or background assumptions related to a hypothesis or research project such that, over time, change in the beliefs of the members of the community occurs; and 4) "equality of intellectual authority": that all members of the community, including women, minorities, and other groups whose voices are often marginalized "should be acknowledged in the public process of critical engagement" (ibid., 76–79). Although questions can be, and have been raised about the feasibility and/or desirability of one or another of these criteria, many feminists view Longino's re-conceptualization of how objectivity can be fostered as a significant advance over those emphasizing that individual scientists should, on their own, engage in critical self-reflection concerning their unexamined background assumptions and/or recognize that their perspectives are partial. Longino went on to further refine her account to address some of the issues raised about her initial presentation of it (Longino 2002, 134).

So, at least as many feminists understand the implications of "situatedness," it does not entail relativism. As we have seen, feminists argue that not all background assumptions are equally warranted; nor, they argue, are all hypotheses. In addition, if, as many feminists (and others) argue, scientific reasoning and knowledge are inherently social, then an emphasis on how science communities can increase the diversity of their members, and encourage critical engagements among its members who may bring different assumptions to their practice, along the lines Longino proposes, are viewed by many feminists as promising approaches to fostering the objectivity of scientific knowledge.

Changes in Primatology Beginning in the 1970s

According to Laura Fedigan and other feminist primatologists, many assumptions about and approaches to gender relations and roles, and gendered aspects of social dynamics that characterized primatology through the 1970s and early 1980s have changed in dramatic ways. Moreover, they have changed among primatologists who do not identify as feminists and/or do not view their research as in any way shaped or motivated by feminism. We discuss the relevant changes, as Fedigan does, using the tool "the remaking of theoretical understandings" that Schiebinger proposes.

The Remaking of Theoretical Understandings

In "The Paradox of Feminist Primatology: The Goddess Discipline?" a title whose source and meaning we discuss in the last section of this chapter, Fedigan cites several arguments that maintain that there have been substantial changes in the theoretical understandings of the field since attention to gender became a research priority beginning in the 1980s.

> Many primatologists have shifted their priorities about the types of research questions that are asked and funded. [Evidence of this] is the increasing study in the 1980s and 1990s of gender-related issues … There are many examples that show that [women primatologists] have created a new vision of the female primate in large part by simply choosing to find out more about her and her relations with the others of her species. (Fedigan 2001, 50)

As we've noted in several contexts, Fedigan and others (e.g., Haraway 1989 and Hrdy 1986) attribute the initial interest to a relatively small group of feminist primatologists in the 1970s. But, Fedigan and others argue, the changes have had a broad impact on the theoretical understandings of other primatologists. Fedigan cites many examples of such changes, three of which we briefly discuss.

In the 1980s and 1990s, studies of prosimian primates that focused on females led to models of female prosimian primate behaviors that were quite different from those based on earlier studies of Old World Monkeys. For example, dominant female lemurs became a focus of research interest, and led to the development of an ecological and evolutionary model "of the conditions under which it is advantageous for female primates to be dominant over the males of their groups" (e.g., Jolly 1984).

In several influential papers, Donald Sade has argued that his long-term studies of rhesus monkeys indicated that the fundamental structuring principle of most Old World monkey societies involves female kinship bonds. As Fedigan describes them, Sade's studies and publications explore the implications of his findings "that affiliative interactions occur mainly among matrilineal kin, that mothers avoid mating with sons, and that groups fission along kinship lines" (Fedigan 2001, 60).

Fedigan also notes Richard Wranghman's modeling of primate structures. Wranghman maintains that females of many primate groups are the first to "distribute themselves" in relation to available food, water, and other resources. This is followed by males distributing themselves "according to the spatial and social pattern of females available" (ibid., 59–60). As Fedigan notes, Wranghman's model was important "in moving us away from a concept of male primates as 'owners' of females." For it suggests that female sociality and patterns of distribution are prior to and consequential for those of males (ibid., 60). In her survey, Fedigan could provide only a brief synopsis of Wranghman's model, but it is worth citing some of Wranghman's specific findings as example of changes in theoretical understandings to which attention to female primates led.

In "An Ecological Model of Female-Bonded Primate Groups" (1980), Wranghman begins by noting that previous analyses of primate groups focused on both sexes together, rather than just females. Wranghman takes up the suggestion that a more promising approach would be to study "the species distribution of individual, rather than group, characteristics" (ibid., 263), and he focuses on groups that are "female-bonded." He defines the phrase this way:

> FB [female-bond] species are defined as those in which females maintain affiliative bonds with other females in their group and normally spend their lives in the group where they are born (or if the group divides, with females who were born in the same group) ... A group is defined as a set of individuals in a closed social network. (ibid., 264)

Wranghman's analysis draws on an extensive body of data yielded by longitudinal studies of many primate species. As he interprets this data, in at least 23 species, restricted movement of females, and regular movement of males between groups, has been confirmed, suggesting that these species are female-bonded. In 26 species in which little was known about intergroup residence patterns, female bonding (including consistent if somewhat differentiated versions) were identified based on "grooming, aiding, huddling, or dominance interactions" (ibid., 266).

Female cooperation (rather than competition) within a group in terms of attaining resources and competing against females in other groups, Wranghman hypothesizes, is largely a function of ecological factors. He further hypothesizes that FB groups "have evolved as alliances to protect access to optimal food patches" (ibid., 270) and he provides details of the observations of a number of FB groups that serve as tests of this hypothesis. There are exceptions, which Wranghman attributes to available food resources, diet, and group size.

Accordingly, he draws the following tentative conclusion.

> Caution is appropriate in interpreting the tests of the model. Given the qualitative nature of the comparisons, the small number of non-FB species, and the lack of competing hypotheses, the fit between theory and data may be more apparent than real. Further and more exacting tests are therefore required. (ibid., 286)

But, Wranghman goes on to note that "The review of ecological and behavioral data shows that many FB species conform to the expectations of the model."

> High-quality food patches are an important focus of ranging patters and intergroup interactions, and females are actively involved in group travel decisions and encounters with other groups ... Females are also less involved in group leadership and interactions in non-FB than in FB species. (ibid., 286)

Attention to female bonding, and the findings that resulted in terms of females' "roles" in social dynamics are quite different from previous studies of primate species that emphasized "male bonding" and "male dominance" as key aspects of what was taken to be "the primate pattern," including assumptions about how males determine the distribution of females.

More Philosophical Issues

Is Primatology a Feminist Science?

Some feminist scientists and science studies scholars have viewed the shifts in terms of approaches to gender research priorities, language use, and models of primate behavior, as both reflecting the engagements of self-identified feminists within the field, and as leading toward the adoption "of the values and practices of feminist science" (e.g., Haraway 1989; Rose 1994; Rosser 1986).

Of course, we need at this point to ask just what would make a science *feminist*. Although there were calls in the early and mid-1980s for developing "feminist science" (e.g., Bleier 1984), many who engage in feminist science studies, including scientists, reject the idea. A more appropriate question, they argue, is "what does it means to do science as a feminist?" and in addressing it, to focus on the different questions, methods, and values that feminist scientists might bring to their research (e.g., Longino 1990). Among the values commonly cited as in keeping with (but not exclusive to) feminism that are related to empirical adequacy of hypotheses and theories are

- "taking the female point of view" (incorporating the study of females into key aspects of research);
- critical self-reflection (engaging in reflection about one's location within historically and culturally specific scientific and social contexts, and how it may affect one's reasoning);
- valuing more inclusive science communities in terms of socially salient categories (gender, race, culture, sexuality, and so forth) that impact experience and perspectives;
- when appropriate (and, some would argue, this is very often the case in the biological sciences), looking for alternatives to reductionist and determinist models (e.g., working to develop pluralistic and interactionist models) of natural phenomena (e.g., Fehr 2001); and
- prioritizing research likely to foster human well-being (e.g., as identified in Fedigan 2001).

Do those who describe primatology as "a goddess discipline" (Rose 1994) or "a genre of feminist theory" (Haraway 1989) have such values in mind and/or are they responding to the kinds of change we have considered in this chapter in terms of increased attention to females – their points of view, actual (complex) behavior, and significant roles in social dynamics? And/or are they responding to the emergence of new questions – such as "what selection pressures led to *smaller size in females*?", a question which serves as an important counterpoint to the longstanding emphasis on the selection pressures favoring larger males? (e.g., Fedigan 2001).

As reviews by other primatologists of Haraway's *Primate Visions* make clear, many primatologists – however much their research focuses on questions that reflect one or more values feminists embrace (such as "taking the female perspective") – vehemently reject the idea that primatology is, or is on its way to becoming, "a feminist science." Fedigan suggests three reasons that her colleagues reject such an idea. First, despite the substantive contributions of researchers who self-identify as feminists, which is itself a relatively small number, feminism is generally perceived by primatologists as "outside" science, not "within" science. In addition, many continue to understand feminism as a purely political and value-laden perspective, and continue to view science as apolitical and value-free. Finally, Fedigan remarks, it may be that some scientists conflate "feminism" with "feminine," and view "a feminine science" as undesirable.

Fedigan's summary of conversations with colleague is instructive in illustrating the sources of resistance to drawing associations between the science they practice and feminism.

> Many primatologists I have talked to say that they changed their practices in order to make science better, not necessarily because feminists thought they should do so but because it was right, scientifically right, to flesh out the picture of female primates, to consider questions from a female, as well as a male, perspective, and to research issues of concern to women as well as men. (ibid., 66)

"This suggests to me," Fedigan continues, "that the goals of feminists and of scientists may sometimes dove-tail – at least the goal of producing a better, more inclusive science, one that incorporates the female perspective of both the primatologists and the animals that they study" (ibid., 66).

Finally, Fedigan argues, it is clear, based on examples such as those we have considered in this chapter, that what is genuinely important are "the myriad ways in which primatologists have become increasingly gender sensitive and gender inclusive" since the 1980s. Few other sciences that have been the subject of feminist critique, she argues, "have moved so quickly, so extensively, and so willingly to rectify the previous androcentric aspects of their practices" (ibid., 48). One might argue that in the interest of gaining insight into relationships between gender and science, recognizing the role of feminist scientists in helping bring about such changes is also important.

We turn next to Developmental Biology, another science that has undergone extensive changes in terms of its approaches to sex/gender.

5   Developmental Biology

Developmental Biology and Developmental Biologists

What processes lead from a single diploid cell, created by the fusion of an ovum and a sperm, to an organism that has many kinds of cells – blood cells, neurons, and muscle cells, among others? How do cells organize themselves into functional structures such as complex organs and tissues of various kinds? What ensures that cell division continues only until it has produced the right number of cells (the number of cells in your legs, for example)?

In the 10th edition of his textbook, *Developmental Biology,* Scott F. Gilbert cites these questions as among those developmental biologists ask about embryonic and fetal development (Gilbert 2014). ("Embryo" refers to an organism in the early stages of growth and differentiation – in humans, this involves the first two months of development; "fetus" refers to an organism during later stages of development.) The developmental process is complex, involving genes and their interactions, hormones, and other factors. Indeed, Gilbert argues, just the notion of an embryo is "staggering."

> Forming an embryo is the hardest thing you will ever do. To become an embryo, you had to build yourself from a single cell. You had to respire before you had lungs, digest before you had a gut, build bones when you were pulpy, and form orderly arrays of neurons before you knew how to think. One of the critical differences between you and a machine is that a machine is never required to function until after it is built. Every animal has to function even as it builds itself. (ibid., 1)

Human fetal development, during which all the essential organs are growing and new features such as hair and eyebrows emerge, is no less amazing or complex.

Developmental biology is an experimental and interdisciplinary science. Its methods and hypotheses are informed by research in other areas of biology as well as areas of biochemistry. Those engaged in research into development are trained in a variety of disciplines, and in recent decades, research in developmental biology has become increasingly integrated with molecular biology, evolutionary biology, and ecology. Such interdisciplinarity notwithstanding, Gilbert points out that each science "is defined by the questions it asks" and that these are a function of its priorities. The research priority of developmental biology is, in Gilbert's words, to provide "a coherent causal network from gene through functional organ" (ibid., 2). Thus, the questions emphasized in developmental biology often differ from those pursued in sciences with which it interacts. Gilbert notes, for example, that to know that XY mammals are usually male and XX mammals are usually female does not *explain* sex determination.

> The developmental biologist wants to know *how* the XX genotype produces a female and *how* the XY genotype produces a male. Similarly, a geneticist might ask how globin genes are transmitted from one generation to the next, and a physiologist might ask about the function of globin proteins in the body. But the developmental biologist asks how it is that the globin genes come to be expressed only in red blood cells, and how these genes become active only at specific times of development. (ibid., 1–2)

In this chapter, we consider some of the ways in which feminist biologists and science scholars have critically engaged assumptions and hypotheses of developmental biology, and some philosophical issues raised in and by these engagements. We begin with a discussion of an account of human sex determination – of how the developing human typically comes to be female or male – accepted during the 1980s. (The qualification reflected in "typically" reflects the fact that there are three kinds of hermaphrodite; and that not all human females have an XX genotype, nor do all human males have an XY genotype.) Of course, embryonic and fetal development involve much more than the emergence of primary and secondary sex characteristics – respectively, internal gonads and external genitalia. But the question of how sex determination occurs has been of longstanding interest, as we saw in our discussion of Aristotle's explanation of it in Chapter 1. More to the point of the present discussion, developmental biologists recognize the phenomenon as a significant developmental milestone. In addition, given that sexual reproduction is central to evolutionary theory, the fact that organisms of many species typically develop into females or males is important to that theory, as well as to biological sciences whose research questions and hypotheses are informed by evolutionary theory.

Historically and currently, accounts of human sex determination have also been understood to reveal a significant role for biology in contributing to (or determining) purported sex and sex/gender differences beyond the anatomical differences related to sexual reproduction. We considered examples of such hypotheses in earlier chapters, including Aristotle's arguments that female inferiority or "incompleteness" is based on how their biology is different from that of males, and Parental Investment Theory's tenet that gametic dimorphism results in sex and sex/gender differences in mating and parenting strategies.

For obvious reasons, what biologists take the implications of sex determination to be for human gender differences is of interest to feminists. But so, too, are the details of the account of sex determination with which we begin. After summarizing it, we consider the critiques feminists have offered of it and the alternative approaches to the study of sex determination they have proposed. We next turn to the account of fertilization that prevailed well into the 1980s, feminist critiques of it, and alternatives that they have proposed. Feminists maintain that the accounts of sex determination and fertilization they criticize had much in common. Both, they argue, associated males and biological entities characteristic of males – for example, the Y chromosome and sperm – with traits long associated with "masculinity," particularly activity and causal efficacy; and females and biological entities characteristics of females – for example, ovaries and eggs – with traits long associated with "femininity," particularly passivity and/or dormancy. Feminists maintain that these associations are stereotypical, and quite often metaphorical. As importantly, they argue, the associations resulted in little if any attention to, or information about, female sex determination, or about the role of the egg and other features of female biology in fertilization, over the course of many decades.

Since the 1990s, there have been significant changes in developmental biology's approaches to and understandings of human sex determination, fertilization, and sex/gender. We conclude with a discussion of some of these changes and consider arguments that they provide insights into relationships between gender and science. In this context, we consider two questions raised by those considering such changes: "Is developmental biology a feminist science?" and/or "Has feminism changed developmental biology?" As we will see, the questions are as controversial within developmental biology as similar questions about primatology are within that field.

Philosophical Issues

As in previous chapters, the research and critiques we consider involve philosophical issues. We consider four such issues. We begin a discussion of "theoretical virtues" that we will continue in Chapter 7 in relation to research in neurobiology. Theoretical virtues, such as empirical adequacy and explanatory power, are features of scientific theories that have been found to be valued by scientists and to inform their practices. Here we consider how the account of sex determination that prevailed into the 1980s might have been viewed to exhibit several such virtues, whatever relationships it might also have borne to then current social beliefs about gender; we consider the reasons why feminist biologists argued that the account did not exhibit some of the virtues in question. We also study feminists' arguments that the account did exhibit the theoretical virtue of simplicity, but that this feature was not appropriate given the nature of the phenomena requiring explanation.

We consider gender's role as an "organizing" or "systematizing principle" in biological research, and feminists' arguments that its role can have a significant impact on a field's research priorities (Schiebinger 1999, 152–155). We find that, despite general agreement that gender often does organize research questions and shape priorities in some biological fields, there are different explanations of why this is the case, and different perspectives on whether its organizing role is warranted.

A third philosophical issue concerns the practice in some areas of biology of not only attributing sex and gender characteristics and differences to *whole organisms*, to which they are at least in principle attributable, but also attributing them to *parts of organisms* – that is attributing stereotypical notions of "masculinity" and "femininity" to biological entities and properties that are not sexed. We discuss feminist arguments that their attribution to entities that are not sexed is inappropriate and consequential in shaping the accounts of sex determination and fertilization they have criticized.

The last philosophical issue we discuss involves feminists' arguments that the "gender dynamics" that characterize a science can impact its research priorities, questions, and hypotheses (Schiebinger 1999). We focus on this issue in our discussion of the changes in how developmental biologists approach sex/gender that have occurred since the late 1980s. We consider disagreements in the field about whether such changes reflect the larger presence of women in the field relative to other biological sciences and/or feminist critiques of the field's earlier priorities and hypotheses. We also consider issues that indicate that the issues underlying these disagreements are quite complex.

Human Sex Determination: The Textbook Account Through the 1980s

In 1985, the account of human sex determination presented in most biology and medical textbooks was consistent with the accounts of embryonic and fetal development in mammals more generally. We focus, as do most textbook accounts of human sex determination, on its different trajectory for individuals with an XX or XY genotype, although we later consider (albeit, briefly) feminist criticisms of the lack of attention to individuals who do not have one or the other of these genotypes.

Some aspects of the textbook account are uncontroversial, although today more is known about the processes involved. Unless otherwise noted, the details in the following summary are currently accepted. During the first five weeks of development, XX and XY embryos exhibit no anatomical differences. Both develop an "indifferent gonad" that includes Wolffian ducts, which in males will develop into the vas deferens, epididymis, and ejaculatory ducts; and Müllerian ducts, which in females will develop into the uterus, cervix, oviducts, and upper vagina. In humans, primary sex determination – the development of gonadal sex – begins in males during the sixth week. Genetic information present on the Y chromosome promotes the synthesis of a protein called the H-Y antigen (referred to in many textbooks as a "male sex hormone," one of a group of such hormones called "androgens"). This synthesis results in the organization of an embryonic testis from the initially indifferent gonad. The embryonic testis contains sperm-producing tubules and it can synthesize hormones. Two of the hormones, testosterone and MIS (Müllerian Inhibiting Substance), further promote development in a male direction. Synthesized testosterone promotes development of the male duct system, and MIS blocks the development of a female duct system.

Until the eighth week, the external genitalia of XX and XY embryos are anatomically undifferentiated. The development of secondary sex characteristics occurs during weeks eight through twelve. Tissues found in both XX and XY fetuses, called "labioscrotal swellings," become a scrotum in males, and labia in females, and the previously bi-potential genital tubercle develops into a penis or a clitoris. Testosterone contributes to the development of the male anatomical features just noted. By week twelve, the structures of male external genitalia are evident in XY fetuses, and female external genitalia are evident in XX fetuses.

As the foregoing indicates, the textbook account provided information about various factors and processes involved in male sex determination. But note that it said almost nothing about female sex determination – about the factors and processes that lead from an indifferent gonad to ovaries, or about the genetic and hormonal factors that result in the development of female external genitalia. This lack of specificity concerning female sex determination was explained as reflecting the fact that female development is "the default trajectory" (imagine what Aristotle would have made of that claim!), and by the assumption that what needs explaining are *exceptions* to that trajectory. Even if the assumption that it is exceptions to a default condition that require explanation is a reasonable view (and there are reasons to think it is not, but they are beyond the scope of the present discussion), feminists maintain that there are two problems with this line of reasoning in the case at hand.

First, even if one assumes that female sex determination comes about because there is something like "a genetic blueprint" for it (a phrase few biologists use today), one would expect scientists to investigate the processes involved. Even more importantly, the assumption that it is exceptions to a default condition that require explanation seems hardly appropriate in the case of sex determination. Roughly *half* of all developing humans do not exhibit the "default" path of development. Both suggest the need for an explanation of female sex determination.

Feminists have argued that the lack of attention to the processes involved in female sex determination was deeply androcentric, and that, rather than an account of "*human* sex determination" as it was called in textbooks, the account was at best an explanation of *male* sex determination. They also suggested that changes to the account of male development might prove necessary when the effects of factors associated with females, such as estrogens, were studied – a prediction, we will see, that was borne out.

And surely one can reasonably ask why an account that focused only on male sex determination was accepted as an account of "human" sex determination. We later consider an answer that some feminist biologists proposed in the 1980s: namely, that its acceptance and that of other androcentric hypotheses during the period reflected politics – and, more specifically, that some biologists perceived the emergence of feminism beginning in the 1970s as a threat to the status quo, as we briefly discussed in Chapter 3. Here we consider an additional possible reason for its acceptance (that is, both may have been factors). The textbook account might have been viewed as exhibiting several "theoretical virtues." Some background about theoretical virtues is in order.

Theoretical Virtues

As noted in Chapter 1 and discussed in Chapter 2, in the 1930s, the physicist Pierre Duhem persuasively argued that there are frequently cases in physics in which the available empirical evidence is not sufficient to establish a theory, or to determine a choice between competing theories. We also noted that the philosopher W. V. Quine later argued that this is characteristic of all sciences. More specifically, Quine argued that there is "empirical slack" between theories or hypotheses and the evidence that supports them. In response to such arguments, as well as to analyses of scientific practice, many scientists and science scholars came to recognize that scientists often make use of one or more "theoretical virtues" in accepting a theory or choosing between competing theories. Here we consider whether, and if so to what extent, the textbook account of sex determination did exhibit, or might have been perceived to exhibit, one or more theoretical virtues.

The account surely did exhibit one theoretical virtue scientists value: that of "simplicity" since it gave a single explanation of sex determination – namely, the presence or absence of a Y chromosome. But simplicity is a virtue only if a theory explains *all* that needs to be explained. The textbook account at most explained male sex determination. The account might also be credited with the virtue of "external consistency" (also known as "conservatism") because it was consistent with general accounts of mammalian sex determination. But, as feminist biologists noted, those general accounts also emphasized the role of the Y chromosome and did not explain female sex determination. Finally, feminists argued that given that the account only explained male sex determination, it failed to exhibit the theoretical virtues of empirical adequacy (accounting for all relevant data), generality of scope (applying to all relevant phenomena), and explanatory power (providing an explanation of all the phenomena in need of explanation). It failed to exhibit these virtues, feminists argued, precisely because it did not explain female sex determination (e.g., Bleier 1984; Fausto-Sterling 1985).

Given these considerations, one might reasonably ask why the problems were not recognized by the scientists who developed or accepted the textbook account? An alternative to the suggestion that only politics motivated its acceptance (mentioned earlier and considered later) presents itself from a contextualist perspective. One can and should note that, at the time the textbook account was taken to be empirically adequate and sufficiently explanatory, androcentrism was not taken to be *problematic*. Indeed, feminists have sought to show how androcentrism reflects longstanding assumptions about sex and sex/gender differences, dating at least to the ancient Greeks, according to which, men and males are taken to be more causally efficacious and significant actors than are women and females, and activities associated with men are taken to be the more consequential. Nor, in the period when the textbook account was developed and accepted, was androcentrism limited to developmental biology or to biology in general. In the 1970s and 1980s, feminists in many sciences provided evidence that androcentrism shaped research questions and priorities in their fields – including archaeology, anthropology, sociology, among other social sciences, as well as psychology, medicine, and medical research. This is not to suggest, let alone to argue, that androcentrism was not and is not a problem, or that it did not or does not result in hypotheses and theories that are at least incomplete and/or flawed in other respects. It is to say that feminist science scholarship itself suggests that androcentrism was pervasive in science; and that its problematic role in various sciences was not recognized until the 1970s and 1980s. So, the failure to recognize that the textbook account did not exhibit some key theoretical virtues might have been at least in part a function of the general failure within many sciences to recognize that androcentrism is problematic. This is also not to say that the presence of androcentrism in science was or is unrelated to sociopolitical beliefs and values involving gender. It is to say that the most interesting cases – the cases that provide insights into science – are those in which scientists did not or do not recognize androcentrism as problematic.

Feminist Critiques and Contributions

We turn now to some additional problems with the textbook account that feminists cited, and the alternative focuses and questions they proposed. Recall that the account maintained that, "in the absence of male factors" of various kinds, the indifferent gonad of the XX embryo develops into the internal female reproductive system (i.e., ovaries); and simply noted that the female structures of external genitalia "become evident" by twelve weeks (Fausto-Sterling 1985, 81). Writing in 1985, feminist biologist Anne Fausto-Sterling argued that the inattention to female sex determination was "odd." "In comparison with all we know about male development," she wrote,

> The view that females develop from mammalian embryos deficient in male hormones seems, oddly enough, to have satisfied the scientific curiosity of [developmental biologists] for some time, for it is the only account one finds in even the most up-to-date texts. (Fausto-Sterling 1985, 81)

Nor, feminists argued, did the assumption that the female line of development is "the default" trajectory of sex determination explain or justify inattention to it. For one thing; as we earlier noted, virtually half of embryos follow the so-called default trajectory and, so, call for an explanation of how that trajectory unfolds. For another, as Fausto-Sterling argued, the lack of attention to female development was not in keeping with the general norms and practices of developmental biology.

> Because it is in the nature of research in developmental biology to always look for underlying causes, failure to probe beyond the "testosterone equals male" – "absence of testosterone equals female" hypothesis is a lapse which is at first difficult to understand. (ibid., 81)

Fausto-Sterling also noted that as early as 1985 it was becoming known that more than androgens were circulating during development. She cited a then recent review article that maintained that the XX gonad begins to synthesize quantities of estrogens about the same time that the XY gonad begins making testosterone. The authors of the review also noted additional circulating hormones.

> Embryogenesis necessarily takes place in a sea of hormones ... derived from the placenta, the maternal circulation, the fetal adrenal glands, the fetal testis, and possibly the fetal ovary itself ... It is possible that ovarian hormones are involved in the growth and maturation of the internal genitalia. (ibid., 81)

Given the emerging information about hormones, feminists argued that there were problems even in the account given of male sex determination. They pointed to the lack of attention to the role of the maternal environment and the possible role of estrogens in male development, pointing out that male and female fetuses produce both hormones (it is the amount that differs). They also criticized the lack of attention to the X chromosome in sex determination. It is true that the Y chromosome promotes the synthesis of the H-Y antigen that in turn promotes the organization of embryonic testes. The testes in turn synthesize testosterone and MIS, which contribute to further development in a male direction. But it is *not* the presence of the Y chromosome that determines the testis' ability to synthesize these hormones. It is genetic information on the X chromosome, as well as on one or more of the twenty-three pairs of non-sex chromosomes, that code for androgens (including testosterone) and estrogens. So, although the Y chromosome is somehow involved in the translation of some of this genetic information, Fausto-Sterling argued that the emphasis on its role and the lack of attention to the other factors just cited, were unwarranted (ibid., 79–85).

Feminist biologists also argued that conclusions could not be drawn from research apparently establishing the effects of prenatal androgens until a similar amount of research is done on the organizing effects of prenatal estrogens. The fetal environment is rich in both, males and females synthesize both (again, it is the amounts that differ), and there are continuous conversions among the three families of hormones (e.g., Bleier 1984; Fausto-Sterling 1985). The difficulties in isolating their effects are illustrated by the reversal of claims that circulating male hormones cause "aggression" in rats. The behaviors so described are now attributed to estradiol (a form of estrogen) converted from androgens by brain cells.

Feminist biologists also criticized the unidirectional causal model of male sex determination, citing experimental results that indicate complex and often nonlinear interactions between cells, and between cells and the maternal environment, during every stage of fetal development (Bleier 1984 and Fausto-Sterling 1985). As Scott F. Gilbert and Karen A. Rader put the point some years later,

> Developmental biology ... teaches us that in the determination of mammalian cell fate, context is critical. Whether a cell becomes a skin cell or a nerve cell, cartilage or muscle, is determined by the other cells it meets. A cell is not intrinsically programmed. (Gilbert and Rader 2001, 92–93)

Lastly, feminists criticized the association of males (and entities associated with them, e.g., the Y chromosome and testosterone) with activity and females (and entities associated with them, e.g., the fetal ovary and estrogens) with passivity. We defer discussion of this line of critique until we consider the long-accepted account of fertilization because the critiques feminists offer of the role and consequences of these associations in the two accounts are quite similar.

The "Classic Account of Fertilization" and the Fairy Tale "Sleeping Beauty"

In "The Energetic Egg," an article written for the lay public and published in 1983, developmental biologists Gerald Schatten and Heidi Schatten argued that there were striking parallels between the Grimm Brothers' fairytale "Sleeping Beauty" and the then accepted account of fertilization. They challenged the account's portrayal of the egg as passive and dormant until "penetrated and activated by a sperm" based on observations they understood to indicate that the egg's role in fertilization was as active as that of sperm. Although the authors did not describe themselves as feminists, an article published in 1988 by The Biology and Gender Study Group (hereafter, BGSG) describe Schatten and Schatten's analysis as feminist-influenced.

Because the tale of Sleeping Beauty represents a straightforward approach to the core critiques feminists offer of this account of fertilization (and, thus, it is frequently cited by feminists), we also use the tale in discussing these critiques. By conducting a decidedly unscientific survey of friends and colleagues my age, I have learned that although most of the women read or were exposed to the fairytale as children, this was not the case for most of the men. So, we begin with a brief synopsis of the tale.

Sleeping Beauty as Told by the Brothers Grimm

In 1812, German academics Jacob and Wilhelm Grimm published what is now known as *Grimms' Fairy Tales*. They were not in fact "the authors" of the tales, which included "Sleeping Beauty," "Snow White," and "Rapunzel." Interested in preserving a rich German oral tradition, they conducted many interviews to collect stories that were part of German folklore to publish them before they were lost.

"Sleeping Beauty" begins with a large gathering at which a kingdom is celebrating the birth of a princess, a gathering to which many fairies are invited. As expected, the fairies gave gifts to the princess, such as beauty, that only fairies can bestow. But one fairy, angry because she was not invited to the celebration (whether by design or mistake), attended anyway and bestowed a "gift" that was more like a curse. She decreed that the princess would prick her finger with a needle at the age of 16, and fall into a deep sleep as would all other living things within the castle grounds. Fortunately, a "good" fairy had not yet presented her gift and was able to modify that of the angry fairy. She decreed that the princess and other living things within the castle grounds would awaken when a prince kissed the princess, and that the two would marry. Devastated by the curse, the King and Queen ordered all needles removed from the kingdom; but one was overlooked and the princess found it, pricked her finger, and she and the rest of the castle's occupants fell into a deep sleep. News of the sleeping princess spread far and wide, and many princes traveled long distances and struggled to get through a dangerous forest that had grown up around the castle in an effort to find the princess. Years would pass before a prince succeeded in gaining access to the castle. He kissed and awakened the princess and their wedding followed.

Schatten and Schatten argued that according to what they called "the classic account of fertilization," eggs like the princess, are dormant, the former until fertilized by sperm, the latter until kissed by a prince and "awakened." The classic account, they argued,

> has emphasized the sperm's performance and relegated to the egg the supporting role of Sleeping Beauty, a dormant bride awaiting her mate's magic kiss which instills the spirit that brings her to life. The egg is central to this drama, to be sure, but it is as passive a character as the Grimm Brothers' princess. (Schatten and Schatten 1983, 29)

Based on their observations using scanning electron microscopy, and on their reevaluation of reports of observations others had earlier made, Schatten and Schatten maintained that "it is becoming clear that the egg is not merely a large yolk-filled sphere into which the sperm burrows to endow new life" (ibid., 29). Rather, they contended, "recent research suggests the almost heretical view that sperm and egg are mutually active partners" (ibid., 29). What Schatten and Schatten observed is that when the egg and sperm interact, the sperm does not "burrow into the egg"; rather the egg "directs" the growth on its surface of small finger-like projections (called "microvilli") to clasp the sperm so that it can draw the sperm into itself. Interestingly, Schatten and Schatten also noted that as early as 1895 E.B. Wilson had published photographs of sea urchin fertilization in which the egg's extension of microvilli to the sperm was visible.

And there was more evidence to come that would support Schatten and Schatten's observations and interpretation of them. In 1991, feminist anthropologist Emily Martin, who also compared the classic account of fertilization to "Sleeping Beauty" and maintained that scientists "had constructed a romance" between egg and sperm "based on stereotypical male-female differences," studied research into fertilization undertaken in the late 1980s and early 1990s in a lab at Johns Hopkins University (Martin 1991). It had long been assumed that sperm used mechanical means to get through the zona (a thick membrane surrounding the ovum) and penetrate the egg. As Martin chronicles, earlier investigations had emphasized "the mechanical force of the sperm's tail" in enabling fertilization (ibid., 492). To their great surprise, Martin states, investigators at Johns Hopkins

> Discovered … that the forward thrust of sperm is extremely weak, which contradicts the assumption that sperm are forceful penetrators. Rather than thrusting forward, the sperm's head was now seen to move back and forth. The sideways motion of the sperm's tail makes the head move sideways with a force that is ten times stronger than its forward movement. In fact, its strongest tendency, by tenfold, is to escape by attempting to pry itself off the egg. (ibid., 493)

Given these observations, Martin noted, the scientists concluded that "the egg traps the sperm and adheres to it so tightly that the sperm's head is forced to lie flat against the surface of the zona." Indeed, they told Martin that it was "like Br'er Rabbit getting more and more stuck to tar baby the more he wriggles" (ibid., 493). But, Martin also noted that the initial papers written by these researchers following these observations continued to suggest that the sperm is the active party "who attacks, binds, penetrates, and enters the egg." The only difference, she argued, is that "sperm were now seen as performing these actions weakly." Indeed, Martin reported that it was a full three years before those involved "re-conceptualized the process" to give the egg a more active role, and stopped portraying sperm as not only penetrating the egg, but also "producing the embryo" (ibid., 494).

At the same time, Martin argued, as did Gilbert in a postscript to the 1988 article he co-authored with members of the BGSG, different but still disturbing gender stereotypes and gendered metaphors entered the literature on fertilization after the egg's active role was recognized (Gilbert 1994). Books written for pregnant women, as well as medical texts, continued to describe the sperm's journey through the female reproductive tract in ways analogous to that made by the princes in "Sleeping Beauty." The female tract, like the woods the princes needed to traverse, was described as inhospitable, indeed at times as dangerous, to sperm. Some texts portrayed the egg as "a femme fatale" who victimizes sperm. And in an example Gilbert cites, biobehaviorist Meredith Small, proposed that "sperm wars" occur in the female reproductive track. Small attributed the competition to the actions or potential actions of the egg: "female philandering" (or at least the possibility thereof), she maintained, is the cause of "sperm wars." Further, she argued that males themselves "evolved as an extension of their brawling sperm" and "the problem of sperm – and thus of males – is, of course, the fault of females" (quoted in Gilbert 1994, 8).

In his textbook, Gilbert remarks that accounts and interpretations of fertilization continue to assume "there's an unfriendly battle going on between sperm and egg". Is there an alternative interpretation of the data now available about fertilization? In *Developmental Biology*, Gilbert states that sperm are capacitated by factors in the female reproductive tract, and that sperm capacitate eggs – that is, their interactions are necessary for fertilization. "Neither the egg nor the sperm is the 'active' or 'passive' partner," he writes. "The sperm is activated by the egg, and the egg is activated by the sperm" (Gilbert 2014, 149).

Feminists' critiques of and alternatives to the accounts of sex determination and fertilization we have considered have much in common. They emphasize androcentrism and the role of gender stereotypes in describing aspects of the processes involved in the two phenomena, particularly the associations of males and features of their biology, on the one hand, with activity and causal efficacy, on the other hand; and the associations of females and features of their biology, on the one hand, with passivity and dormancy, on the other hand. In addition, feminists argue that both accounts invoke gendered metaphors: they attribute characteristics stereotypically associated with "masculinity" or "femininity" to biological entities that are not sexed. As we noted in earlier chapters and will discuss in the next section, one way to understand feminists' concerns with the role of gender stereotypes and gendered-metaphors is that they carry normative as well as empirical content – that is, they are evaluatively thick rather than purely factual.

More Philosophical Issues

Gender as an Organizing or structuring principle in biology

We have discussed feminists' arguments about the role of gender stereotypes in describing biological processes and behavior, about the role of gendered metaphors in describing entities and processes that are not actually sexed, and about biological concepts and hypotheses presented as if they describe purely factual phenomena that feminists argue are, in fact, not purely factual but evaluatively thick. One approach to such issues focuses on how gender can operate as an organizing or structuring principle in biology. We explore this approach because it illuminates differences between non-feminist and feminist approaches to the role of gender in biology.

In an argument representative of others feminists offer, Londa Schiebinger maintains that gender often functions in biology as a framing principle – a basic organizing or structuring principle that contributes to, or in some cases determines, the research priorities of a biological field (Schiebinger 1999, 152–155). It serves, in other words, to identify a "path" for research to follow. Schiebinger and other feminists point out that there are many "paths" scientists can take in studying some phenomenon or, indeed, in deciding what phenomena to study. Sometimes, which path they follow is a function of factors beyond their control and might not, in fact, be their preferred path. For example, most researchers are not self-funded. They work for universities, or for privately or publicly funded research labs, or for private corporations. Institutional priorities and funding opportunities may require following a path that is not what a researcher would have chosen. And some approaches or paths are determined by the core tenets, and experimental and theoretical developments, that characterize the field within which a scientist works. The path research takes is also influenced, to one or another extent, by background assumptions of varying kinds, sometimes recognized and sometimes not, by those whose research is to some extent informed by them.

Feminists argue that gender's role as an organizing principle is complex. We find, for example, that some biologists consciously embrace and defend gender's role as an organizing principle in biological research. At the same time, feminists argue, as Schiebinger makes the point, that gender "can and does sometimes function as *a silent organizer* of scientific theories and practices" (ibid. 152; emphasis added). Such arguments, which may point to unquestioned background assumptions, intend to highlight just how pervasive the concept of gender is and how it can apparently seem to be so obvious that its role in guiding research is neither questioned nor scrutinized.

In what follows, we consider three explanations of gender's role as an organizing or structuring principle in biology: 1) arguments offered by some feminist biologists in the 1970s and through the mid-1980s that its role was primarily a function of gender politics; 2) arguments, on the part of many biologists, that its role as an organizing or structuring principle is purely empirical; and 3) arguments offered in the 1990s by feminist philosophers and scientists that its role reflects *both* scientific *and* social contexts.

Arguments offered by feminists from the 1970s through the mid-1980s often maintained that gender's role in biology is primarily the result of gender politics – and, specifically, in an interest during the period in undermining feminists' challenges to the idea that divisions in power and roles between men and women are the inevitable consequence of biology. In 1984, for example, neurophysiologist Ruth Bleier argued that gender is of "overriding importance"

> in the patriarchal civilizations that have been our cultural context for the past several thousand years ... a particular, consistent, and profound bias shapes [biological theories], theories about women in particular, and scientific explanations of the perceived social and cultural differences between women and men. (Bleier 1984, 2)

Not only has such bias been a pervasive aspect of patriarchal culture, Bleier argued, but challenges levelled by feminists seeking "to break out of positions of subordinance" led some biologists to redouble their efforts to establish that the status quo is a function of biology (ibid., 2). Bleier's arguments are representative of some others offered during the period.

It might be suggested that the fact that concepts of sex and gender have been used to organize and structure research devoted to other species undercuts the claim that the emphasis placed on them always reflects an interest to explain and justify differences in women's and men's roles and power. But arguably this is not always the case. We have considered, for example, how "the primate pattern" proposed in the mid-twentieth century that emphasized the functionality and universality of male dominance, was attributed to humans as well as many nonhuman primates. In Chapter 7, we consider feminists' analyses of research in empirical psychology and endocrinology focusing on sex differences in rodents that were taken as evidence that, in mammals, there are cross-species' and *universal* sex differences in behavior and temperament.

But we have also noted that a scientist might not recognize that the warrant for the assumptions about gender that guides her or his research, like all research assumptions, need to be scrutinized. Nor can we assume that all biologists for whom gender functions as an organizing principle recognize that their hypotheses carry implications for social beliefs and policies. We discussed such issues when we introduced the concept of "background assumptions" in Chapter 1. These considerations suggest that how and why gender has an organizing role in a biological science or research program might not be solely or primarily a function of concerns about changing gender relations.

We turn now to a different explanation of gender's role as an organizing or structuring principle that is offered by some biologists and evolutionary theorists. In some biological sciences, gender is by no means a *silent* organizing principle, i.e., unrecognized as systematizing research questions and priorities. Many view gender's role as an organizing principle in evolutionary theorizing to be dictated by the theory itself. If one assumes, as evolutionary theorist Theodosius Dobzhansky made the point, that "nothing in biology makes sense except in light of evolution" – a view with which many if not most biologists agree – then attention to traits "selected for" because they are or were once conducive to reproductive success is obviously a research priority. The mating and/or parenting strategies of females and males, including whether they are different, obviously warrant study.

Consider, for example, philosopher Michael Ruse's argument focusing on the role of sex/ gender in evolutionary theorizing about humans:

> Evolutionary biologists – Darwinians in particular – are absolutely obsessed with sex and have had much to say on the subject. What would you expect from a theory that puts reproduction right up at the top of things needed for evolutionary change? (Ruse 2012, 186)

Ruse's arguments about the appropriate role of sex/gender in organizing research emphasize sexual selection theory and, specifically, Parental Invest-ment Theory. While acknowledging that the relationship between sexual selection theory and the specific sociopolitical contexts in which it was initially proposed and subsequently developed, are "complex," Ruse main-tains that attention to sex/gender and to sex/gender differences are, in the end, rooted in the biology of evolution.

> Or put it this way, if in looking at evolutionary questions you are not going to talk about reproduction and the fact that it is females who have the offspring, then what are you going to talk about? (ibid., 192.).

> You may not like the talk of aggressive males and coy females, but the point is that unless these strategies work, they will be dropped. And this leads straight into the reason why Darwinian biologists are going to persist in using such terms, or alternative language denoting the same things. It is the baby-bearing business ... the fact is that it is the females who have the babies. They are the limiting resource. And especially when you start to get up the scale to ... organisms like mammals and birds, this counts. (ibid., 194)

Of course, as feminists note, it is not the case that females take full responsibility for gestation in all sexually reproducing species (e.g., in some species males sit on eggs until they are ready to hatch). But Ruse is discussing humans and human evolution. His general argument is that attention to sex/gender is unavoidable given the importance of reproduction to evolutionary change; and that if one accepts sexual selection, the greater investment of females and male competition for access to females, are in the end biological facts – even if the language used to describe them has often been informed by culturally specific gender stereotypes and gendered metaphors.

By way of anticipating a third explanation of gender's role as an organizing principle, it is worthwhile recalling the fundamental challenges to the traditional account of parental investment to which Ruse is appealing, as well as the challenges raised by research undertaken by primatologist Sara Blaffer Hrdy considered in Chapter 4. Hrdy's observations as well as those of other feminist primatologists indicate that females of many primate species are far from coy: they initiate sexual encounters, engage in fierce competition with other females, and manipulate males in a variety of ways to ensure their own reproductive success. As we saw, Hrdy doesn't disavow Parental Investment Theory. Rather, she takes females' greater investment in offspring to lead to what Haraway described as behaviors that are aptly described as "motherhood with a vengeance."

So, too, the attribution of "aggressiveness" or "aggression" to males, which Ruse mentions, is arguably not as straightforward as he suggests. In Chapter 3 we began a discussion of feminist critiques of the assumptions about "male aggression" that take it to result from a male's relatively small investment in offspring and because females are "a limited resource" as hypothesized in Human Sociobiology. We noted that feminists argue that scientists in evolutionary theory and other biological sciences who routinely attribute "aggression" to the males of one or more species fail to provide clear identity criteria for the concept. They also challenge the assumption that "aggression" is appropriately extrapolated from one species to others. In what ways, they ask, are mounting behaviors exhibited by male rodents "aggressive"? And even if they are, what do they have in common with behaviors exhibited by male primates, including human males, that are categorized as "aggressive'?

But a third explanation warrants consideration: namely, that the role of gender as an organizing or structuring principle in many biological fields reflects *both* specific scientific contexts *and* specific social contexts. From this perspective, one can and should ask whether the descriptions of males and females, or hypotheses about sex and sex/gender differences, *are* purely factual or empirical, as Ruse and many scientists assume. One need not deny that the focus on sex and sex/gender, and specific hypotheses proposed about them, are often based on *scientific* theories (e.g., evolution) to argue that *social* beliefs, values, and assumptions *also* substantively contribute to their content. In Chapter 2, we considered historians' accounts of how aspects of Darwin's scientific and social contexts informed his reasoning. In Chapter 3, we considered feminists' arguments that Parental Investment Theory assumes gender stereotypes, gendered metaphors, and evaluatively thick concepts (concepts that carry both empirical and normative content). We also saw, in Chapter 4, how a very different picture of the behaviors of female primates taken to result from their greater investment in offspring emerged in the work of primatologists such as Jeanne Altmann, Sara Blaffer Hrdy, and Frans de Waal, when gender stereotypes and androcentrism were avoided. And we have seen in this chapter that quite different accounts of fertilization are possible when androcentrism, gender stereotypes, and gendered metaphors are avoided.

This is not to say, indeed it would be flatly inaccurate to claim, that the alternatives feminists propose are themselves purely empirical or value-free. Like the priorities and hypotheses they are intended to replace, they too are informed by both empirical and normative content. The point is that, arguably at least, the third explanation – that gender's role as an organizing principle in biology reflects both current scientific and social contexts – is more warranted than the first, which maintains that its role is primarily or solely a function of gender politics, and more warranted than the second, according to which the questions and hypotheses about sex and gender that scientists propose are purely empirical and do not reflect historical social contexts.

As we discussed in Chapter 4 and will continue to discuss in what follows, there are biologists and philosophers who find the emphasis the third explanation places on social and historical contexts, both internal and external to science, to be inappropriate.

Gender Dynamics in Developmental Biology

Women entered developmental biology, as they entered primatology, earlier than they entered other biological sciences, and they have succeeded in it as indicated by their research productivity and the positions they have attained. Some argue that their presence and perspectives have led to new questions, research priorities, and knowledge in the field. In an extended analysis, Scott F. Gilbert and Karen A. Rader focus on "the historical intersection of late-twentieth century feminism with developmental biology." They argue that "the knowledge critiques" that resulted from that intersection came to transform the field (Gilbert and Rader 2001, 75). In 1988, The Biology and

Gender Study Group at Swarthmore College, of which Gilbert was the founding member, also attributed a significant role to that what they called the "feminist critique" of developmental biology.

> We have come to look at feminist critique as we would any other experimental control. Whenever one performs an experiment, one sets up all the controls one can think of in order to make as certain as possible that the result obtained does not come from any other source. One asks oneself what assumptions one is making. Have I assumed the temperature to be constant? Have I assumed that the pH doesn't change over the time of the reaction? (BGSG 1988, 61)

Feminist critique, they argued, asks "if there may be some assumption that we haven't checked" that reflect gender bias. "In this way," they continued, "feminist critique should be part of normative science. Like any control, it seeks to provide critical rigor, and to ignore this critique is to ignore a possible source of error" (ibid., 61).

It is important to note that this section began by discussing aspects of the *gender* dynamics of developmental biology, and specifically the relatively larger presence of *women* in the field compared with other fields. Their participation, Gilbert and Rader argue, has made a difference to the field. But when we began to anticipate some of their arguments about what brought about these changes, and those offered by The Biology and Gender Study Group, we switched gears, so to speak, because these analyses emphasize *feminism* – suggesting that women (and men) responsible for changes in developmental biology's assumptions and priorities were or are feminists or feminist friendly. But this is not true of all those whose work is compatible with feminists' goals. We will also find that many women in developmental biology strongly resist the idea that their gender impacts their research, reject the characterization of their work as "feminist," and reject the claim that feminism brought about the changes that have occurred developmental biology.

Gilbert and Rader recognize these complexities; and cite examples that they take to demonstrate that the relationships between gender, feminism, and developmental biology are complex, in need of further study, and controversial.

> Like any important and anomalous observation in science, the apparent success of women in developmental biology suggests more questions than it answers ... What constituted the success of women developmental biologists

and how did it come about? ... Another, more complicated question follows from this line of inquiry: namely, have the number and achievements of women in developmental biology during [the period from the 1930s to the present] made a difference? Have these women made developmental biology a "feminist science" – or has feminism changed the means by which we do developmental biology in other ways? (ibid., 75)

We have so far considered critiques offered by feminists of a once-accepted textbook accounts of sex determination and "the classic account" of fertilization, feminists' arguments for the need to study female sex determination, and Schatten and Schatten's observations of and arguments that the egg has an active role in fertilization. We now turn to additional critiques of and contributions to developmental biology that Gilbert and Rader credit with leading to changes in the approaches to sex/gender in the field. Their analysis includes many more examples than we can discuss, and is often more nuanced than our discussion reflects. But the summary provided below reflects the basic lines of argument they offer for the view that further study of relationships between gender, feminism, and the changes that have come about in developmental biology, is likely to yield important insights into gender and science.

Gilbert and Rader note that there were earlier critiques of the language used in developmental biology than those we have so far considered, and that they were offered by men as well as by women. In 1976, the Women's Caucus of the Society for Developmental Biology published the pamphlet, *Sexism Satirized*, a collection of cartoons that satirized sexist content in then-current medical and biology textbooks. The preface stated that the pamphlet "was made possible through generous contributions of material of SDB members of both genders." One cartoon satirically portrayed the Y chromosome as a large weightlifter, adorned with a sash that read "Master Y," holding up a small woman whose feet were bound, to satirize the following statement in what was, at the time, a well-respected textbook.

> In all systems that we have considered, maleness means mastery; the Y chromosome over the X, the medulla over the cortex, androgen over estrogen. So physiologically speaking, there is no justification for believing in the equality of the sexes; *Vive la différence!* (Short 1972, Book 2, 70)

The preface to *Sexism Satirized* includes the statement, "Hopefully the authors quoted here will be persuaded to reassess their objectivity in future publications and the awareness of scientists in general will be somewhat heightened" (Doane 1976). Among the things that Gilbert and Rader describe as "remarkable" about the pamphlet is that it was an early and internal critique, as well as the fact that the group that created it included men as well as women. Nor, it should by now be clear, is language *qua* language the primary focus of the pamphlet or of feminist critiques of biological approaches to gender. Referring to maleness as "mastery," portraying the egg as a passive participant in fertilization, and failing to investigate the processes involved in female sex determination, all raise questions about the *empirical content* of the hypotheses and theories in question.

A second example cited by Gilbert and Rader is the research undertaken and arguments offered by Schatten and Schatten that contributed to the initial recognition of the egg's role in fertilization. As the BGSG notes, these researchers did not describe themselves as feminists or describe their work as motivated by feminists' concerns about the imposition of gender stereotypes and gendered metaphors. However, as we earlier noted, the study group describes their investigations and findings, as it does the research we next discuss, as "feminist-influenced."

Gilbert and Rader also describe arguments offered by molecular biologists Eva Eicher and Linda Washburn, as influenced by feminism and as contributing to changes in how gender is approached in studies of development. In 1986, Eicher and Washburn criticized then current accounts of sex determination, particularly the assumption that development in a female direction is the "default" trajectory. This assumption, they argued, was only possible "if one confused primary and secondary sex determination" (ibid., 90). Here is one example they used to support this claim:

> If you castrate a mammalian embryo, its phenotype becomes female. But that is secondary sex determination, and has nothing to do with whether the bipotential gonad rudiment becomes a testis or an ovary. Primary sex determination is actually a bifurcating path, and both testis and ovary formation are active, gene-directed events. (ibid., 90)

In terms of primary vs. secondary determination, recall that it is not genes on the Y chromosome that determine the testis' ability to synthesize testosterone and MIS, syntheses that allow for the development of male secondary sex characteristics. Rather, it is genetic information on the X chromosome, as well as on one or more of the twenty-three pairs of non-sex chromosomes, that allows for this ability. Eicher and Washburn maintain that because of the conflation of secondary sex determination with primary sex determination, sex determination became conflated with "male sex determination."

Eicher and Washburn proposed that the evidence of genetic involvement in testis development makes it highly probable that the development of ovarian tissue results from an equally active and genetically determined process. They argued that studies of the genotypes of XX individuals who do not develop ovaries might well provide information about the genes involved in normal female development (Eicher and Washburn 1986).

In light of the cases just summarized, the Biology and Gender Study Group argued that

> A feminist critique ... involves being open to different interpretations of one's data and having the ability to ask questions that would not have occurred within the traditional context ... [These researchers] have controlled for rather than let the ancient myth run uncontrolled through their interpretations. (BGSG 1988, 67–68)

Gilbert and Rader do note that Schatten and Schatten, and Eicher and Washburn, might not agree with the characterization of their work as "feminist." Perhaps what they and members of BGSG have in mind is that the avoidance of gender stereotypes, gendered metaphors, and androcentrism by these researchers, is fundamentally compatible with the approaches taken by those who describe their own approaches as feminist.

Has Feminism Changed Developmental Biology?

What should we make of the complex relationships between gender, feminism, and developmental biology discussed in the previous section? At the very least it seems that one needs to acknowledge, as Gilbert and Rader make the point, that the study of relationships between gender, feminism, and developmental biology in the twentieth and twenty-first centuries raises "as many questions as it answers," and warrant additional study. And in this section, we consider some of the complex issues we have mentioned, but not yet discussed in any detail, to which they are referring.

Gilbert and Rader, The Biology and Gender Study Group, and Gilbert's 1994 postscript to the BGSG 1988 analysis, all point to significant changes in developmental biology that are fully commensurate with feminists' interest in and concerns about research in the field. Gilbert and Rader's conclusions are representative. "We believe," they state, "that feminism has indeed made a difference in developmental biology in several ways." They begin by noting changes in the gender dynamics of the field.

> First, large numbers of women have not only entered the field but have become its exemplars both scientifically and professionally ... [and changes brought about by feminism's challenges to how hiring and promotion are done] have been incorporated in developmental biology in many prominent ways. (Gilbert and Rader, 92)

Gilbert and Rader also cite changes in the research priorities of the field and in the *content* of the theories and hypotheses generated in it, and maintain that "the vocabulary of the field has been transformed, resulting in a less sexist, less culturally biased, and more scientifically congruent view of the world" (ibid., 92). They credit such changes with leading to new approaches to and understandings of fertilization, sex determination, and brain development. By 1990, for example, two genes related to the development of ovaries had been discovered, and a review article in 2007 maintained that it is now known that "absence of testicular hormones does not result in a normal female phenotype; ovarian genes and hormones are necessary" (Blecher and Erickson 2007).

In the end, noting that much more work needs to be undertaken, Gilbert and Rader's "preliminary reply" to the questions, "Have women made developmental biology 'a feminist science'?" and, "Has feminism changed the way we do developmental biology in other ways?" is "a qualified yes" (Gilbert and Rader 2001, 91).

Yet, an analysis undertaken by biophysicist Evelyn Fox Keller of attitudes focusing on how developmental biologists would answer these questions, indicates that not all agree with Gilbert and Rader that women and feminism have brought about the changes in developmental biology they cite. (Gilbert and Rader also cite Keller's analysis and the questions it raises.) Keller notes that not all developmental biologists credit feminism with the changes that have occurred in the field, let alone agree that, in some sense, developmental biology is "a feminist science" (Keller 1997). Many see the changes in developmental biology as resulting from maturation from within, and as apolitical – rather than brought about by feminism, which many understand to be *outside science* and inherently political. These views are similar to those we encountered among some primatologists who resist the idea that feminism is responsible for changes in their science. In neither field is there denial that changes in assumptions about, approaches to, or knowledge concerning gender, have occurred. What is contested is how the changes have come about.

Keller finds that there are other sources of the resistance to crediting feminism with changes that have occurred in developmental biology. One is the perception among practitioners in the field (both women and men) that feminist scientists and science scholars "advocate a 'feminine' science"; the other is the perception "of hostility in the feminist literature to science itself" (ibid., 16). To those who equate feminism and femininity, Keller observes, a "feminine science" is associated with "intuitiveness" and "softness," despite, as she notes, many feminist arguments that such associations are incorrect. Keller notes that these connotations are worrisome to women in developmental biology who "fear that such a notion will be used to discredit either them or their work." Indeed, she suggests that this fear is often "so acute that any question about the role of gender in science is … seen as likely to be counter to their interests as scientists" (ibid., 16).

Suppose one is unaware of the developments within and evolution of feminist science scholarship over the last four decades, and assumes that it continues to reflect hostility towards science as some, *but only some*, early feminist engagements did. Suppose one also does not consider the degree to which feminist science scholarship is engaged in *by scientists*, or discounts the role of feminism in their practices. Then one's resistance to any proposed relationships between feminism and changes in developmental biology is understandable – even if, as Keller notes, it frustrates feminist scientists and science scholars.

Perhaps a conclusion parallel to that offered by Linda Fedigan about relationships between feminism and changes in primatology is also appropriate in terms of developmental biology. Recall that Fedigan took the emphasis on evidence in feminists' critiques, and in the insistence among primatologists that the changes in their field were driven by evidence, as indicating that feminist scientists, and scientists more generally, are motivated by the same goal: "the goal of producing a better, more inclusive science, one that incorporates the female perspective of both the primatologists and the animals that they study" (Fedigan 2001, 66).

Recall, as well, Fedigan's conclusion about what is really important about the changes in primatology. She argued that "the myriad ways in which primatologists have become increasingly gender sensitive and gender inclusive" and their efforts "to rectify the previous androcentric aspects of their practices" are, in the end, of most importance (ibid., 66).

Could we say that the same is true of the changes that have occurred in developmental biology? Or if we are interested in fuller understandings of the relationships between gender and science, is it equally important to recognize the role of gender dynamics and feminists' contributions in bringing about the changes that occurred in developmental biology?

6   Medicine

Introduction

Generations before, indeed centuries before, there were medical schools and medical degrees as we know them, there were "physicians," "doctors," and "medical experts," just as there were "physicists" and "biologists" before there were graduate degrees in those fields. Hippocrates (460 BCE–370 BCE), an Athenian, is considered the father of modern medicine. The Greek philosopher Galen (130 CE–216 CE) was a physician and surgeon, and served as the personal physician of several Roman emperors. He wrote extensively on anatomy, physiology, and other medical topics, and his writings influenced European medicine well into the eighteenth century.

In this chapter, we consider medical theories about biological processes and organs that are unique to women, and how, during specific periods, medical professionals argued that some or all of them are detrimental to women's health, if not pathological – and how, during other periods, they were *not* so regarded. Our discussion is limited to Western medicine and focuses on three time periods. The first and longest spans the period from ancient Greece through much of eighteenth century Europe. The second period is American and English medicine during the nineteenth century and early years of the twentieth. The third period is American medicine from the 1960s to the present.

In broad strokes, medicine of the first period took many aspects of women's and men's biology to be quite similar. A heart is a heart, whether it is a man's heart or a woman's heart. The ways in which the anatomy and biological processes of men and women obviously do differ – sex organs and biological processes involving reproduction – were held to be, although different, analogous. (It was, however, a common belief during this period that men's and women's brains are different in important ways.) In stark contrast, medicine in the nineteenth century and the first decades of the twentieth maintained that men's and women's biology are fundamentally different. Medical experts viewed the causes of illnesses in men and women, including those both were subject to, as having different origins, with women's originating in their sex organs and men's in the environment. They also maintained that, given their reproductive organs and processes, women needed more extensive and frequent medical treatments than men do. Our discussion of medicine in the twentieth and twenty-first centuries illustrates that many of the sexist views common in nineteenth-century medicine gradually gave way to a more balanced view of biological processes unique to women, although menstruation and menopause continued to be viewed negatively in medical and physiological textbooks well into the late 1980s, and pregnancy and childbirth became increasingly "medicalized." We also consider how medical assumptions about women's biology impacted research studies and clinical trials in the 1980s and 1990s in ways that were androcentric and problematic in other respects. Finally, we briefly consider the relatively new field of Evolutionary Medicine, and several hypotheses offered by those working in it about the adaptive significance of menstruation. The hypotheses we consider approach and explain menstruation in more positive terms than has traditional medicine.

Before beginning our discussion of approaches to women's health during these periods, two caveats are in order. In focusing on Western medicine's assumptions about and approaches to women's biology and health, we are excluding vast areas of medical history and practice. It is not only Western women who have been profoundly, and not always positively, affected by medical views and practices. Women in poor and developing countries have had to contend with the added burdens of poverty, unsanitary conditions, and less effective treatment of the health issues that affect them. There is an extensive literature on the conditions these women face, and on the efforts undertaken by organizations such as the World Health Organization to identify and work to address them. The issues involved are too large and complex for us to consider, but they are important and pressing.

The second caveat is this. The medical theories and practices, and critiques thereof, that we consider, concern issues involving *sex/gender and Western medicine*, not Western medicine *tout court*. There are, of course, analyses and critiques of other aspects of medical practice, indeed critiques of medicine in general, but they are not germane here. Western medicine has provided many significant benefits in terms of human health, including the eradication of many diseases and treatments for many others that have prolonged and saved lives. Physicians, nurses, and other medical professionals who participate in organizations such as "Doctors without Borders" often risk their lives to provide badly needed medical treatment; and many physicians and other medical professionals provide free care to poor patients in numerous clinics. We focus on what feminists argue have been problems in Western medicine's approaches to women's health and health issues, and the relationships that feminists claim to have identified between such problems and specific historical and social contexts.

Philosophical Issues

The discussions included in this chapter involve philosophical issues, most of which have been introduced in earlier chapters. We briefly summarize them here.

Contextualism

We have adopted a contextualist approach to scientific research in previous chapters. Recall that this involves studying relationships between scientific theorizing, on the one hand, and then contemporary scientific and social contexts, on the other hand. In considering the medical research and practices of the historical periods on which we focus, we again take a contextualist approach. As in previous chapters, this includes attention to how scientific language can be evaluatively thick, reflecting contemporary sociopolitical contexts – medical language in the case at hand. And we pay special attention to how scientific language carries ontological commitments: that is, commitments to there being specific objects, events, and processes. In the case of medical theories and assumptions about women's bodies and health, commitments to the existence of specific illnesses and diseases often have concrete and substantive consequences. For example, a nineteenth century theory that diseased ovaries cause many illnesses, including "personality disorders," led to what we now know to have been the unnecessary removal of the ovaries of many women.

Representative Sampling

We discussed representative sampling in Chapter 4 and the extrapolation of results obtained during the study of one group to others. These issues loom large in our discussion of biomedicine, where we find that large research studies and clinical trials of drugs and surgical procedures frequently excluded women as participants, even when the illnesses or conditions studied affects women as well as men. The results of such studies and trials were often extrapolated to women, sometimes with significant and unfortunate consequences.

Research Priorities

We also consider research priorities in medicine, and feminists' arguments that there are women's health issues that warrant greater priority than they have often received. For too long, feminists argue, insufficient attention was paid to diseases such as breast cancer that primarily affect women, and to women suffering from diseases taken to only or primarily affect men, such as cardiovascular disease. We also briefly discuss differences between the research priorities of traditional medicine in relation to women's reproductive biology and the research priorities of Evolutionary Medicine.

Has Feminism Changed Medicine?

In the concluding section, we consider the questions, "has feminism changed medicine?" "And, if so, to what extent?" "And what more," do feminists argue, "needs to change?" We find, as we did in terms of other sciences we have considered, that answers to these questions are complex.

Finally, there are philosophical issues concerning medicine that we will be unable to consider because the literature devoted to them is more extensive than we can engage. One of the more important involves philosophical analyses and disagreements about how "disease" and "health" should be defined and understood.

Medicine and Sex/Gender: From Ancient Greece Through the Late Eighteenth Century

Physicians, anatomists, and physiologists from ancient Greece through much of the eighteenth century, found basic similarities between men's and women's anatomy and organs. Men's and women's livers, hearts, and many other organs, it was argued, are basically the same. (As we have noted, this was not taken to hold of brains. Men's brains and women's brains were held to be different in several respects.)

Physicians and anatomists also described structural analogies between organs and biological processes that are obviously different in men and women – notably, those related to reproduction. There were features of women's and men's reproductive biology that were unknown for much of the period (for example, ovaries were not discovered until the seventeenth century). But from the second century CE when Galen proposed analogies between men's and women's genitalia – writing that "Turn outward the women's, turn inward, so to speak, and fold double the men's, and you will find the same in every respect" – until well into the seventeenth century, descriptions and illustrations of the vagina and uterus portrayed them as analogous to the penis and scrotum. Figure 6.1 depicting the uterus and vagina as analogous to the penis and scrotum, is attributed to the seventeenth

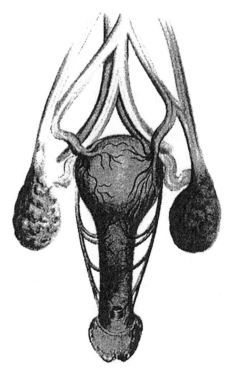

Figure 6.1 Seventeenth century depiction of the uterus and vagina. Vidius 1611, Vol. 3. Photo taken from Weindler 1908: 140.

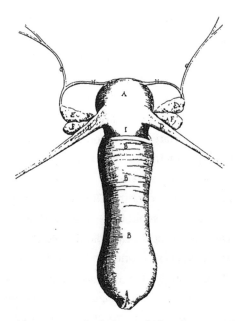

Figure 6.2 Sixteenth century depiction of female reproductive organs as phallus-like.
Attributed by Weindler 1908:141 to Bartisch's *Kunstbuclze,* 1575 (MS Dresdens. C. 291). Photo taken from Weindler 1908, fig. 104b, p. 144.

century Italian anatomist, Vidus Vidius. Figure 6.2, an illustration of female external genitalia that depicts them as phallus-like, is attributed to the sixteenth century German anatomist, Georg Bartisch.

The physiological analogies assumed to obtain between women and men were based on then-current understandings of the human body, its relationships to environments, and what is required to maintain human health. What was required, it was believed, were continuous therapeutic processes taken to originate in the body itself. Anthropologist Emily Martin cites the account given by historian Charles E. Rosenberg of this theory. The body, Rosenberg maintains, was viewed as necessarily engaging in "physiological adjustment" to maintain health and life.

> The body was seen, metaphorically, as a system of dynamic interactions with its environment. Health or disease resulted from a cumulative interaction between constitutional endowment and a circumstance. One could not well live without food and air and water; one had to live in a particular climate,

subject one's body to its particular style of life and work. Each of these factors implied a necessary and continuing physiological adjustment. The body was always in a state of becoming – and thus always in jeopardy. (Rosenberg 1979, quoted in Martin 2001, 30)

In addition, as Martin describes Rosenberg's account, auxiliary assumptions resulted in other factors being understood to "govern this interaction." One, as Rosenberg describes it, is that "every part of the body was related inevitably and inextricably with every other." Another is that "the body was taken to be a system of intake and outtake – a system which had, necessarily, to remain in balance if the individual was to remain healthy" (quoted in Martin 2001, 30).

The assumption of fundamental similarities, and in some cases analogies, between aspects of women's and men's biology, together with the view of the body as engaging in necessary adjustments to maintain health and life, contributed to the belief that even processes unique to women, such as menstruation, had analogies in processes thought to be more common in men. For example, Galen and others who accepted these views understood menstruation as a process that rids a woman's body of excess blood – a process that men could bring about through other means, such as bloodletting. It was also common to view the "flush" associated with menopause as analogous to men's sweating – both serving as "safety-valves" that rid the body of excess fluid thought to be harmful to a body's overall balance. Some physicians also believed that menstruation, the flushes associated with menopause, and sweating on the part of men, functioned to rid the body of impurities and, as such, maintain health.

This is not to say that women were considered equal to men during these centuries. Humans were thought to be superior to other species because they possessed more "heat" – that is, differences in heat were understood to be related to differing degrees of perfection. Like Aristotle, many physicians and biologists took what they believed to be women's lesser amount of heat as the basis of and evidence for their biological inferiority. And like Aristotle, students of medicine and psychology argued that women are intellectually inferior to men, and that their emotions, rather than their rational faculties, are dominant. Hippocrates is generally assumed to have first posited an illness he termed "hysteria," and viewed to be a psychological and behavioral syndrome that affects only women. The term "hysteria" comes from the

Greek "hysterikos" (of the womb, suffering in the womb), and the affliction was thought to be the result of disorders of the womb that cause women to exhibit excessive and volatile emotions and behavior.

The differences cited in the previous paragraph notwithstanding, over the course of many centuries physicians, anatomists, physiologists, and biologists claimed to find more similarities between women's and men's anatomical structures, biological organs, and physiology – both those related and unrelated to reproduction – than they did differences. And in terms of the recognition of analogies between men's and women's genitalia, from the perspective of current theory about sex differentiation during development that we discussed in Chapter 5, the views of Galen and others about analogies between men's and women's external genitalia were prescient. Recall that until the eighth week, the genitalia of XX and XY embryos are anatomically undifferentiated. The development of genitalia occurs during weeks eight through twelve. At that time, tissues found in both XX and XY fetuses, called "labioscrotal swellings," become a scrotum in males, and labia in females, and the previously bi-potential genital tubercle develops into a penis or a clitoris.

As we explore in the next section, the emphases on similarities and analogies between aspects of women's and men's biology would come to be abandoned, indeed strongly denied, by physicians and psychologists in nineteenth and early twentieth-century America and England.

Medicine and Sex/Gender in Nineteenth and Early Twentieth Century America and England

The changes the nineteenth century brought to medical views about women's biology and health were dramatic and multifaceted. First, differences between men's and women's organs and biological processes related to reproduction came to be viewed as significant and of consequence in relation to their "roles" and health. Second, women's organs and biological processes related to reproduction came to be understood as detrimental to women's health, if not pathological. Writing in 1900 about the "ravages" thought to be wrought by the "sexual storms" of female puberty, menstruation, pregnancy, and childbirth. The president of the American Gynecology Society used an analogy with shipwrecks to describe the dangers women encounter during each phase of their reproductive years. His observations of the effects on women of biological processes unique to them were common during the period and shaped by specific theories we consider in this section. And the accounts of each such process, we will see, were evaluatively thick; they carried both normative and empirical content that had important impacts on women's medical treatment.

> Many a young life is battered and forever crippled on the breakers of puberty;
> if it crosses these unharmed and is not dashed to pieces on the rock of
> childbirth, it may still ground on the ever-recurring shallows of menstruation,
> and lastly upon the final bar of the menopause ere protection is found
> in the unruffled waters of the harbor beyond reach of sexual storms.
> (Dr. Engelmann, quoted in Hall 1905, 588)

Third, medical experts came to view women's reproductive organs, her ovaries and uterus, as the source of disease and illness involving other organs, and as dictating that her role is that of wife and mother. Of course, this view of women's proper role was not new. What were new were the ideas that it is women's ovaries and uteruses that determine that role, and that their proper or improper functioning determines women's health or illness in every respect – even in terms of illnesses and diseases to which men are also subject. As one medical professor declared in 1870, it is "as if the Almighty in creating the female sex, *had taken the uterus and built up a woman around it*" (M.L. Holbrook, emphasis in original; quoted in Ehrenreich and English 2005, 132). Many physicians and psychologists emphasized women's ovaries as dictating all aspects of femininity, including women's psychology (ibid., 131). One physician, speaking of women's ovaries and uteruses, declared "Women's reproductive organs are pre-eminent."

> They exercise a controlling influence upon her entire system . . . They are the
> source of her peculiarities, the centre of her sympathies, and the seat of her
> diseases. Everything that is peculiar to her, springs from her sexual
> organization. (John Wilbank, quoted in Smith-Rosenberg 1973, 59)

Nowhere is the shift from earlier views that some of men's and women's biological processes, although different, serve analogous functions, to the view that some biological processes unique to women are detrimental, more obvious than in nineteenth century medical views about menstruation. So, too, puberty in young women came to be viewed as a biological process that, if not properly overseen and managed, would result in women's reproductive organs failing to develop properly.

Puberty and Menstruation

Given the dominance attributed to a woman's ovaries and uterus through their reproductive years, it was taken to be of paramount importance that these organs develop normally during puberty. Because medical professionals at the time accepted the theory that human bodies possess a limited amount of energy, they warned young women that, during puberty, energy must not be diverted from the ovaries and uterus as this would compromise their development. As feminist historian Caroll Smith-Rosenberg describes physicians' recommendations,

> At the commencement of puberty, a girl should curtail all activities. One doctor advised the young woman to take to her bed from the first signs of a discharge until menstruation was firmly established, months or perhaps years later. Not all doctors took so extreme a position, but most did warn that girls should not engage in any absorbing projects at this time. (ibid., 62)

"Indeed," Smith-Rosenberg goes on to note, "physicians routinely used the energy theory to sanction against ... education, factory work ... indeed virtually any interests outside the home" (ibid., 62). Girls were advised to engage in routine domestic tasks such as cooking and bed-making, and to avoid strong emotions. Physicians, Smith-Rosenberg argues, created a stereotype of the pubescent girl, a vision of that girl as "a fragile and ethereal creature," who had to cope with constant traumas (ibid., 61). In contrast, she notes that Michael Ryan, a physician writing in 1843, maintained that during puberty a young man is vigorously developing the physical and mental capacities that would allow him to "perform the noble pursuits assigned him by nature," again assuming a gender stereotypes (ibid., 61).

We have seen that for many centuries menstruation was viewed as analogous to the kinds of purging of excess fluids and/or impurities that men purged by sweating or blood-letting. But, from the perspective of nineteenth and early twentieth century medicine, menstruation was *a unique kind* of biological process, a process that was debilitating if not pathological – putting women at high risk from the onset of menses until menopause. Women (at least those belonging to the middle or upper class), medical experts maintained, require extensive rest before and during their monthly periods. In 1870, the zoologist, Walter Heape, expressed a common view when he described menstruation as a "severe, devastating, periodic action ... [that leaves behind] a ragged wreck of tissue, torn glands, ruptured vessels, jagged edges of stroma, and masses of blood corpuscles." Not surprisingly, Heape argued that menstruation requires medical treatment. "It would seem hardly possible," he maintained, that a woman could "heal satisfactorily without the aid of surgical treatment" (quoted in Laqeuer 1987, 31–32). Popular books advising women about menstruation sounded similar cautions.

It is quite likely, of course, that women in the nineteenth century, like many if not most women, experienced some degree of cramping, water-weight gain, and other discomforts in the days before and during menstruation, and that some experienced mood changes to some degree or other. But what explains the view among nineteenth century physicians that menstruation is an *illness* – a fundamentally evaluatively thick conclusion? We return to this question in a section in which we consider feminists' critiques and contributions.

Pregnancy and Childbirth

If puberty and menstruation posed threats to women's health that required they severely limit their activities, pregnancy and childbirth were viewed to be even more hazardous and more in need of medical supervision – at least in the case of middle and upper-class women. During the entire course of pregnancy, medical experts maintained, women must confine themselves to home and limit their activities to relatively simple domestic chores such as supervising household staff. Women were also told that to avoid damage to the fetus, pregnant women had to avoid certain kinds of "prenatal impressions."

> All "shocking, painful or unbeautiful sights," intellectual stimulation, angry
> or lustful thoughts, and even her husband's alcohol and tobacco-laden
> breath – lest the baby be deformed or stunted in the womb. (Ehrenreich and
> English 2005, 122)

Physicians also prescribed a long period of convalescence following birth, reflecting their view of "the pathological nature of childbirth itself" (ibid., 122). Indeed, in the nineteenth century there were reasons women needed to convalesce following childbirth. Women who gave birth often suffered from irreparable tears to their pelvis and/or a slipped uterus (prolapse). Giving birth during the period, we learn from women's journals and correspondence, in part due to the pain and potential complications associated with it and in part due to the warnings of physicians, was viewed as a frightening process. As we later discuss, some middle and upper-class women asked physicians for contraceptives (for example, early versions of diaphragms and spermicides) or an abortion. Such requests, for reasons we later consider that were unrelated to religious objections, were vehemently opposed by physicians and social theorists. In the end, many middle and upper-class women in nineteenth century America and England did practice "confinement," limiting their activities to the domestic sphere. (We note their socioeconomic status because, as we later discuss, complete rest during menstruation, and confinement during pregnancy and following birth, were not possible for poor and working class women. As importantly, we learn that most physicians did not view them as necessary for these women, and we consider the reasons they offered to support this view and the alternative explanations offered by feminist and other historians.)

The perceived and actual dangers associated with pregnancy and childbirth might also have been factors leading to the fierce opposition to midwifery among nineteenth century physicians. For centuries, midwives had provided care to women during pregnancy and during and after childbirth. But physicians in England and America undertook substantive and successful efforts to discredit midwives and prevent them from providing such care – again, at least to women of the middle and upper classes. Physicians cited the lack of "medical training" and knowledge on the part of midwives, and many argued that women were incapable of engaging in medicine. However, as we later consider, historians also cite interest among the growing number of physicians and medical experts treating women in promoting the scientific status of *their* practices at a time when science was highly respected as motivating their opposition to midwifery.

The potential complications of pregnancy and birth for women in the nineteenth century cannot and should not be underestimated. We have already noted that some women experienced tears to their pelvis and/or prolapse. And the first general study devoted to American births, undertaken in 1915, reported that for every 10,000 live babies born, 61 women died. By the late 1970s, the number of maternal deaths for every 10,000 births, had dropped to two. One question feminists raise is whether together with the relative lack of knowledge in nineteenth century medicine compared to that of obstetricians and gynecologists today, the sheer *number* and *frequency* of pregnancies experienced by many women in the nineteenth century, contributed to the complications often brought about by childbirth.

Treatments for "Diseased" Ovaries and Uteruses

We have discussed the limited intellectual and physical activities that physicians insisted were necessary for girls going through puberty, and for women during menstruation, pregnancy, and following childbirth. We have also noted that the ovaries and uterus were understood to be the source of many illnesses women exhibited, including headaches, indigestion, irritability, and backaches, as well as diseases in other organs. The medical treatments used to "fix" diseased ovaries and uteruses during this period were often drastic. Historian Ann Douglas-Wood describes what she calls "the local treatments undertaken in stages" during the first half of the nineteenth century, noting that not all women undergoing "local treatments" were subjected to all of them. During the first stage, leeches were placed directly on the vulva or on the neck of the uterus (with care taken not to drop them in a woman's body). In a subsequent stage, various substances were injected into the uterus, including milk, linseed tea, and substances derived from marshmallows. Finally, without anesthesia, but perhaps being given a little opium or alcohol, women were cauterized – either, as Douglas-Wood describes,

> [t]through the application of nitrate or silver, or, in cases of more severe infection, through the use of much stronger hydrate of potassa, or even the "actual cautery" [using] "a white-hot iron instrument." (Douglas-Wood, quoted in Ehrenreich and English 2005, 136)

In the second half of the century, what Douglas-Wood describes as "these fumbling experiments with the female interior," gave way to surgeries, many of which were undertaken to deal with diseases in other organs as well as "female personality disorders" (ibid., 136). For a brief period in the 1860s, some physicians treated "nymphomania" and "intractable masturbation" by removing the clitoris, although many physicians disapproved of the surgery (ibid., 136). The most common surgery for women's diseases, including "personality disorders," was removal of the ovaries. Ehrenreich and English note that in 1906, a leading American gynecological surgeon estimated that 150,000 women in the country had had their ovaries removed (ibid., 136). And they note,

> Since the ovaries controlled the personality, they must be responsible for any psychological disorders; conversely, psychological disorders were a sure sign of ovarian disease. Ergo, the organs must be removed. (ibid., 136)

From the later years of the nineteenth century through several decades of the twentieth, hysterectomies became more common than "female castration." Ehrenreich and English argue that, during that period, "mature women would be treated indiscriminately to hysterectomies" to treat the physical illnesses and behavioral problems that had come to be attributed to the uterus (ibid., 154).

"Invalidism"

"Invalidism," as the condition was called by physicians, struck many middle and upper-class women in England and the United States during the second half of the nineteenth century. Women who suffered from invalidism described themselves, and were described by their husbands and physicians, as suffering from a range of "symptoms" that left them physically and mentally incapacitated. The diagnostic labels experts used to describe the condition included "neurasthenia" (today described in medical dictionaries as "an ill-defined medical condition" associated with "emotional disturbance"). Physicians described the symptoms of invalidism as including weakness, headaches, depression, an inability to concentrate, confusion, anxiety, anger, menstrual difficulties, and "a general debility requiring constant rest" (ibid., 114). Women diagnosed with invalidism were often described as also exhibiting "hysteria." The most respected expert on female invalidism, the American physician S. Weir Mitchell, who took care of hundreds of women apparently afflicted with the condition, described his patients this way:

> The woman grows pale and thin, eats little, or if she eats does not profit by it. Everything wearies her – to sew, to write, to read, to walk – and by and by the sofa or the bed is her only comfort. Every effort is paid for dearly. (ibid., 114)

Mitchell insisted, and other medical experts treating women so diagnosed accepted his view, that invalidism could only be cured through what he called "the rest cure" – complete cessation of intellectual and physical activities, and isolation. We return to invalidism and Mitchell's "rest cure" in the section devoted to feminist critiques.

Menopause

One might expect, given the debilitating effects, or even pathological nature, of menstruation assumed in nineteenth century medicine, that medical experts would understand menopause, as Engelmann did in the passage earlier cited, as "protection [that] is found in the unruffled waters of the harbor beyond reach of sexual storms." But this view was uncommon. Most physicians held negative views of menopause, often referring to it as "the death of the woman in the woman." Like menstruation, it was taken to constitute a medical crisis in terms of both female biology and female psychology. Although the dominance of the ovaries and/or uterus over women's brains, bodies, and psychology ceased when ovulation did, the cessation was argued to lead to other diseases, depression, and "inappropriate behavior." Interestingly, although aging women and men are both more vulnerable to illness and disease, including cancers, it was only in women that the increased incidence of diseases such as cancer was attributed to menopause rather than aging. Moreover, as Smith-Rosenberg found in her extensive study of then current medical texts, physicians argued that "the most significant cause of women's menopausal disease" was the failure during their reproductive years to think and behave in accordance with "the physiological and social laws dictated by the ovarian system"; such violations included contraception, abortion, education, and "undue sexual indulgence" (Smith-Rosenberg 1973, 65).

Feminist Critiques and Contributions

We now turn to questions feminists and other historians raise about the theories about women's biology and health issues that dominated nineteenth and early twentieth century medicine. What explains the emphasis on women's frailty, the "ravages" wrought by their reproductive organs and processes, and the insistence that women suffered from special illnesses and diseases that were related to the organs and biological processes unique to them? Or that illnesses and diseases that men were also subject to were, in the case of women, caused by their reproductive organs rather than the environmental conditions taken to cause the same disease in men?

Histories of nineteenth and early twentieth century medicine, including but not limited to those feminist historians and social scientists have offered, suggest (forgive the cliché) that "a perfect storm" resulted from the convergence of different, but often interrelated, interests and anxieties then prevalent in England and America. We can only focus on some they identify.

But before discussing some of the "non-medical factors" historians cite as impacting how women's health was viewed and treated, we again remind ourselves that nineteenth-century women did experience more complications related to pregnancy and birth than do women today, or at least than women of the middle and upper classes do today. Keeping this in mind, medical efforts to gain knowledge about the processes, and to manage the care of women during them, appear appropriate. One question that warrants consideration is whether such complications and dangers were the only motivation for a view of pregnancy and childbirth, as well as of puberty, menstruation, and menopause, as diseases or even pathological?

Female physicians and activists during the period, and today, argue that there are compelling reasons for not accepting that only interests in safeguarding women's health were factors leading to the diagnoses, recommendations, and treatments of what were taken to be "female illnesses." For one thing, the recommendations for how women should behave during menstruation, pregnancy, and childbirth ignored the fact that many poor and working class women were not able to take time off from their jobs before and during menstruation, or during pregnancy, or following childbirth (Ehrenreich and English 2005, 122). In addition, feminists note that the requests made during the period by female physicians (there were some, but relatively few) and activists that specialists in women's health extend such care to poor and working class women during pregnancy and childbirth, were not successful. This, despite the fact, that the many of the health issues such women faced were as significant as those physicians emphasized and treated in wealthy women, and that poor and working class women lacked the economic resources, as well as the assistance provided by household staff and visiting physicians, that wealthy women enjoyed. The anarchist Emma Goldman, for example, who was trained as a midwife, described "the fierce, blind struggle of the women of the poor against frequent pregnancies." She also noted the suffering of the children of such women, describing them as "sickly and undernourished – if they survived infancy at all" (quoted in Ehrenreich and English 2005, 124).

And surely it is reasonable to assume that poor and working class women experienced at least as much by way of complications during pregnancy and childbirth as did wealthy women. But many physicians maintained that poor and working class women did not need the treatment and assistance that middle and upper class women required. Poor and working class women, it was argued, unlike their wealthy counterparts, were "robust" and fully capable of doing well during pregnancy and birth. In the last third of the century, physicians' arguments for this view came to reflect their understandings of the implications of Darwinism. Ehrenreich and English note that "Civilization," was taken to explain why "the middle-class woman [was] sickly; her physical frailty went hand-in-white glove-hand with her superior modesty, refinement, and sensitivity." In contrast, working class women "were robust, just as they were supposedly 'coarse' and immodest" (ibid., 125). Assuming Darwin's general characterization of the significant sex/gender differences that evolution produced because of the greater selection pressures to which men were subject, it seemed, as feminists describe the issue, that in terms of the wealthy classes, that some came to believe that "men evolve and women devolve" (ibid., 127). Interests in maintaining the current "social order," as well as economics practices, also appear to have been factors in how poor and working class women were viewed. After all, *someone* had to do the work of scrubbing floors and other physical tasks; so, viewing poor and working class women as "robust," functioned to support the social order.

Feminists also cite growing interest on the part of middle class women in pursuing education and careers, and opposition against changes to sex/gender roles, as factors contributing to the perception of what might otherwise be viewed as normal and natural processes, such as puberty and menstruating, as requiring that women avoid intellectual activities. Arguments offered by medical experts against admitting women to college frequently cited the dangers to their reproductive organs. Among the most influential were arguments offered by Dr. Edward H. Clarke in his book, *Sex in Education: Or a Fair Chance for Girls*. Clarke, a professor at Harvard, published the book in 1873 when pressure to admit women to that institution was at its height. As Ehrenreich and English chronicle, Clarke was opposed to women's admission. He appealed to what were common medical views, warning that women who engaged in strenuous mental activity – who studied in a "boy's way" – risked atrophy of their ovaries and uteruses, insanity, and sterility. Studying would cause their brains to drain energy and/or blood from their reproductive organs (ibid., 138–139). Clarke's arguments persuaded many, even some successful women who came to believe education had harmed them; and some colleges cut back on the number of courses women could take in a year (ibid., 141–143). A warning issued by R.R. Coleman, M.D. to women entering colleges or seeking to be admitted was representative. "Women beware," he wrote, "you are on the brink of destruction ... Science pronounces that the woman who studies is lost" (ibid., 141). Clearly, at least some were using medical theories to reinforce notions about women's proper role and to prevent them from entering spheres that men had dominated.

In terms of the complications often attendant to pregnancy and child-birth, as we earlier noted, feminists also suggest that it is likely that frequent pregnancies and births were among the factors contributing to them. As Ehrenreich and English note, given that a married woman in the nineteenth century, without the benefit of contraception or access to abortion, "could expect to face the risk of childbirth repeatedly through her fertile years," and the complications we noted in our earlier discussion of childbirth, it is likely that the maternal mortality rate was even higher in the mid-nineteenth century than the rate reported in 1915.

Feminist and other historians also point to how anxieties about higher birth rates among the poor, immigrants, and other races, compared with those among the white middle and upper classes, contributed to fierce opposition to contraceptives and abortion on the part of wealthy white women. One professor described what he called "the cause of alarm" about birth rates as "the declining birth rate among the best elements of the population, while it continues to increase among the poorer elements" (Edwin Conlin, quoted in Ehrenreich and English 2005, 148). White middle and upper class women were reminded of their maternal duties, based on their sex and race;" and given opposition to their limiting pregnancies or terminating them, many were unable to control the number of pregnancies and births they experienced. There was, as Ehrenreich and English point out, no small amount of contradiction between the insistence that middle and upper class women are frail and frequently suffer from "invalidism," on the one hand, and calls for such women to reproduce sufficiently to insure the number of white middle and upper class babies increases, on the other hand.

Finally, what explains the so-called epidemic of "invalidism" among middle and upper class women – something that itself calls for explanation, and the need for which becomes even more obvious given the absence of "invalidism" (or its perceived absence) among poor and working class women? Class and its conceptual as well as concrete intersections with sex/gender in nineteenth century capitalist economies, feminists argue, provides at least a partial explanation. As earlier noted, feminists point to the importance attributed to men in the middle and upper classes of having wives who could devote all their attention to domestic issues. Writing in 1934, historian Thorstein Veblen described the "lady" of the nineteenth century. Summarizing his account, Ehrenreich and English note that "A 'lady' had one other function [other than that of producing offspring] ... And that was to do precisely *nothing*, that is nothing of any economic or social consequence" (ibid., 116; emphasis in original). Noting that to achieve the pallor and other indicators of their delicate health, some middle and upper class women took to drinking small amounts of arsenic, the authors continue, "for many women, to various degrees, sickness became a part of life, even a way of life ... It was stylish to retire to bed with 'sick headaches,' 'nerves,' and various unmentionable 'female troubles.'"(ibid., 119).

Finally, in terms of an explanation focusing on the expectations attendant to being a "lady," feminists point out that how women in the middle and upper classes were expected to dress and did, must have contributed to the aches, fatigue, and physical disorders they experienced. A corset could exert as much as 20 lbs. of pressure on a woman's torso; and bustles, heavy petticoats, and other garments suspended from their waists, could together weigh over 20 pounds. Female physicians of the time and current feminists argue that the frequency of fainting and of distorted and displaced organs – including prolapse (descent of the uterus sometimes to the point where it exited the body through the vagina) reported during the period, might well be partially explained by the weight of such garments.

So far, we have focused on factors that lend support to an explanation of invalidism based on economic status, women's preoccupation with their health and particularly their nerves (encouraged by their physicians), and even the ways they dressed. But feminists' analyses suggest that, for at least some women, the symptoms of invalidism – including fatigue, loss of appetite, lethargy, and depression – were a function of their unhappiness with the limitations placed on them. The experiences of the activist Charlotte Perkins Gilman are instructive. In her case, one she did not view to be unusual, she eventually came to attribute her bout with "invalidism" to such unhappiness. Gilman reported that her symptoms of "invalidism" were so severe that she sought help from S. Weir Mitchell, who as earlier noted, was regarded as the expert on invalidism and how to treat it. Gilman met Mitchell in the 1880s, after she sent him a letter. Gilman reported that Mitchell ignored the information she provided about only feeling at ease when she was away from home, and demanded that she obey his medical prescription for "a rest cure," which she described as.

> Live as domestic a life as possible. Have your child with you all the while. Lie down an hour after each meal. Have but two hours intellectual life a day. And never touch pen, brush, or pencil as long as you live.

For months, Gilman followed "the rest cure." She would later describe what happened during the period: "[I] came perilously close to losing my mind." Coming to eventually suspect that it was confinement, both during the rest cure and in her home life, that was causing her symptoms, Gilmore left her husband and moved to California with her baby, where, she notes, she picked up brush, pen, and pencil, and became an activist. Her short story, *The Yellow Wallpaper*, published in 1892, is a semi-fictionalized account of the "madness" she experienced during "the rest cure" and of how she came to recognize what she took to be the cause of her symptoms of "invalidism."

As feminist historians point out, we cannot know if what Gilman took to be the cause of her own bout with invalidism applied to many other women, although there is anecdotal evidence suggesting that Gilman's case was not unique. We also cannot dismiss the possibility that some, perhaps even many, wealthy women enjoyed aspects of their life attributed to their "frailty," as the practice among some of them of drinking arsenic to enhance their symptoms suggests. But the foregoing discussion of arguments offered by feminist and other historians would strongly suggest that "female invalidism" in the nineteenth century was not simply or even primarily caused by problems inherent to women's biology.

Sex/Gender and Medicine in the Twentieth and Twenty-First Centuries

We begin our discussion of medicine and sex/gender in this and the previous century focusing on developments relevant to medical understandings of and approaches to women's biology and health during the first 85 or so years of the twentieth century. The good news is that by the end of this period, menstruation, pregnancy, and childbirth were no longer viewed as "illnesses" or pathological. (We will see, however, that Premenstrual Syndrome would later become a much discussed and debated "condition," and menopause would come to be defined as "estrogen-deficiency" leading to Hormone Replacement Therapy, at least early versions of which led to health problems.) And, eventually, ovaries and uteruses were no longer viewed as the source of virtually every illness and disease from which a woman suffered – although as we later discuss, feminists and many gynecologists argue that many of the hysterectomies performed during the previous and current century were unnecessary.

On the other hand, during the first 75 or so years of the twentieth century, pregnancy and childbirth were further "medicalized": management of both by medical professionals increased and labor and delivery came to routinely occur in hospitals, with medical professionals making most of the decisions about how to treat pain associated with them, as well as decisions about whether surgical or other interventions were needed. We defer discussion of the treatment of pregnancy and childbirth until we turn to the emergence of the Women's Health Movement in the 1960s and 1970s that led to changes in how pregnancy and childbirth were viewed and treated. We begin by focusing on how, through the late 1980s, most general physiology texts, as well as most texts written for medical students, described menstruation and menopause.

Menstruation

Although menstruation was no longer viewed as detrimental to women's health (or pathological), it was most often described in physiology and medical textbooks in negative terms. Emily Martin attributes this to the common practice of using metaphors involving "production" to describe menstruation (Martin 2001, 46–51). Menstruation was taken to result from failure: the failure of an ovum and sperm to fuse and produce a zygote. Continuing the "production" metaphor, valuable resources had been devoted to preparing the uterus for implantation and these are wasted when implantation does not occur. The result, as one 1984 physiology text describes it, is "catastrophic disintegration" of the lining of the uterus. That disintegration involves the breakdown of the endometrium, the mucous membrane lining the uterus that develops after ovulation to support implantation, when conception does not occur (ibid., 48). The text described what it called the "catastrophic disintegration" as resulting from the "sudden loss" of estrogen and progesterone that occurs when conception does not, and explains the kinds of disintegration that occur.

> The sudden loss ... causes the blood vessels of the endometrium to become spastic so that blood flow to the surface layers of the endometrium almost ceases. As a result, much of the endometrial tissue dies and sloughs in the uterine cavity ... The sloughed endometrial tissue plus the blood and much serous [blood serum] exudate from the denuded uterine surface, all together called the *menstruum*, is gradually expelled by intermittent contractions of the uterine muscle for about 3 to 5 days. This process is called *menstruation*. (Guyton 1984, 624; emphasis in original)

Martin argues that the terms "ceasing," "dying," "losing," "denuded," and "expelled" are not *neutral*. Rather, she maintains, they suggest "disintegration," "failure," and "dissolution" (ibid., 48). It is true that when conception does not occur, the mucous membrane lining the uterus begins to dissolve and is eventually expelled. But Martin and other feminists argue that the terms used to describe this process have an inappropriate negative connotation (Tavris 1992).

To illustrate this, and to counter arguments that menstruation really *is* "a process of breakdown and degeneration," Martin points to a biological process that parallels what occurs during menstruation, but is not described in negative terms. Our stomach lining is replaced every four days and this is typically described in medical texts in terms of "rebuilding" and "renewal," rather than "degeneration." Martin cites a representative account of the processes involved included in a general physiology text.

> The surface of the stomach must be exceptionally well protected at all times against its own digestion. This function is performed mainly by mucus that is secreted in great abundance in all parts of the stomach. (Guyton 1984, as quoted in Martin 2001, 50)

But, Martin points out, what this and many general physiology textbooks that are not written for medical students do not mention, is the physiological process by which the protection of the surface is accomplished, a process she maintains is analogous to menstruation. The layers of mucus cells need to be "continually sloughed off and digested," just as much of the endometrial tissue dies and sloughs into the uterine cavity in menstruation. Only textbooks written for medical students provide this detail about the processes that protect the stomach lining. Martin cites another physiological process analogous to menstruation that is not described in negative terms. Although it is known among researchers who study male ejaculation that a very large proportion of ejaculate "is composed of shedded cellular material," Martin points out that textbooks in physiology "make no mention of a shedding process let alone processes of deterioration and repair" that occur in the male reproductive tract (ibid., 51).

Martin suggests that the differences in how menstruation and the process that protects the stomach lining are generally described reflect unnecessarily negative views of menstruation.

> One can choose to look at what happens to the lining of stomachs and uteruses as breakdown and decay needing repair, or positively as continual production and replacement. Of these two sides of the same coin, stomachs, which women and men have, fall on the positive side [in many textbooks]; uteruses, which only women have, fall on the negative. (ibid., 50)

Finally, although textbooks of the period commonly describe menstruation in terms that suggest it represents failure because an ovum has been "wasted," Martin notes that often the same textbooks fail to use terms suggesting "loss" or "failure" when noting that, on average, only "one out of every 100 billion sperm" a man has (the average man produces about 500 billion sperm cells during his lifetime), succeeds in fusing with an egg. Women have only somewhere between 300 to 400 ovulated eggs during their reproductive years. Surely, Martin notes, when we consider how many sperm never succeed in fusing with an egg but deteriorate and die, "here is something really worth crying about!" (ibid., 48).

Menopause

Well into the late 1980s, textbooks described menopause negatively, again partly due to the assumption of concepts and metaphors related to "production." The longstanding view of menopause as reflecting a change in how the balance of "intake" and "outtake" was maintained came to be replaced in the mid- to late twentieth century by the view that menopause represents "degeneration." Beginning in the 1960s, menopause was taken to signal the start of the stage of life in which a woman's reproductive organs "failed to produce" sufficient estrogens, a failure that was assumed to be detrimental to her health. Martin chronicles the relevant changes in view by comparing the 1940s and 1950s editions of one gynecology textbook's accounts of the effects of menopause, to the account of menopause in the 1965 edition of that textbook. In the earlier editions, menopause "was described as usually not entailing 'any very profound alteration in the women's life current'" (Novak et al.'s *Textbook of Gynecology*, quoted in Martin 2001, 51). In contrast, in the 1965 edition the authors note that "In the past few years there has been a radical change in viewpoint ... some would regard the menopause as a possible pathological state rather than a physiological one and discuss therapeutic prevention rather than the amelioration of symptoms" (ibid., 51).

The "radical change" spoken of in the 1965 edition reflected a new emphasis on menopause as a state in which a woman's ovaries fail to produce estrogen. As Martin notes, by 1981 the World Health Organization defined menopause as "an estrogen-deficiency disease." A 1986 textbook described menopause as the period during which "the cycles cease and the female sex hormones diminish rapidly to almost none at all." The author went on to describe both the causes of such diminishment and its "negative" effects.

> The cause of the menopause is the "burning out" of the ovaries ... Estrogens are produced in subcritical quantities for a short period after the menopause, but over a few years, as the final remaining primordial follicles become atretic, the production of estrogens by the ovaries falls almost to zero ... At [this] time a woman must readjust her life from one that has been physiologically stimulated by estrogen and progesterone production to one devoid of those hormones. (Guyton's *Textbook of Medical Physiology* 1986, quoted in Martin 2001, 51)

It is true that at menopause the ovaries produce much less estrogen than they did before, but it does not follow that this change has only negative effects on a woman's health. But describing menopause as caused by ovaries "burning out" and postmenopausal women as "devoid" of the stimulation provided by estrogen and progesterone does suggest that menstruation is a negative condition.

To illustrate that it need not be so viewed, Martin notes that not all textbooks of the period described menopause in purely negative terms. Here is a passage from one such text:

> It would seem that although menopausal women do have an estrogen milieu which is lower than necessary for *reproductive* function, it is not negligible or absent but is perhaps satisfactory for *maintenance* of *support tissues*. The menopause could then be regarded as a physiologic phenomenon which is protective in nature – protective from undesirable reproduction and the associated growth stimuli. (Jones and Jones' *Novak's Textbook of Gynecology*, 10th edition, 1981; quoted in Martin 2001, 51; emphasis added)

As earlier noted, we later discuss how the emphasis on the negative effects of so-called estrogen-deficiency in menopausal women led to the development of Hormonal Replacement Therapy (HRT), debates about it, and some of the unfortunate consequences of the initial forms of such therapy.

We turn now to two important developments that have had an impact on medical approaches to and understandings of women's health. One was the emergence in the United States of the Women's Health Movement in the 1960s and 1970s, a movement that became global. The other is the emergence of, and emphasis placed on, biomedical research in the United States beginning in the 1980s and 1990s. We follow these sections with a brief discussion of approaches to women's reproductive health in the relatively new field of Evolutionary Medicine, and how it may lead to changes in how processes like menstruation are understood.

The Women's Health Movement

The beginning of The Women's Health Movement in the United States is marked by the publications of books such as *Our Bodies, Ourselves* in 1973 by what was then called The Boston Women's Health Book Collective, and Barbara Ehrenreich and Diedre English's *For Her Own Good: Two Centuries of the Experts' Advice to Women*, first published in 1978, sections of the 2005 edition we have cited; the founding, across the United States, of numerous groups and organizations devoted to changing women's health care; and the opening, by the late 1970s, of over 1,200 "women's self-help centers." The Women's Health Movement has used publications, grass-root activism, and various organizations, to educate women about their bodies and health, and to challenge the medical practices and treatments of women's health that were viewed as failing women. The movement provided information about menstruation, birth control, abortion, childbearing, menopause, sexually transmitted diseases, and other topics relevant to women's bodies and health. *Our Bodies, Ourselves*, among the movement's most influential publications, is now in its ninth edition and has been translated into many languages. In a relatively short time, the somewhat different goals and demands of those participating in the Women's Health Movement converged and were summarized as "a demand for improved health care for all women and an end to sexism in the health system" (Marieskind 1975, 219).

The changes in women's health care the movement brought about or contributed to are many and various. We focus on several that are representative and important. By the mid to late 1980s many hospitals had changed protocols for labor and childbirth, abandoning those that not only isolated women during the processes but were also experienced by women as "dehumanizing." Labor wards and operating rooms were replaced in many hospitals by facilities that reflected "a family-centered approach" to birth, allowing (indeed, encouraging) husbands and other family members to stay with a woman during labor, and husbands to attend and take part in birth. Many hospitals started offering classes preparing women and their partners for childbirth and postnatal care, including classes preparing interested couples for "natural childbirth," and many women became able to choose the medications and other treatments provided during labor and birth. By the late 1980s, some hospitals had established "birthing centers," home-like facilities in which women delivered their babies often assisted by a midwife. Of course, many such opportunities for managing their labor and delivery were often limited to women who could afford them. And it is also true that "medicalization" of pregnancy has continued, as reflected in the use of ultrasounds to monitor development, and in the availability of prenatal genetic screening and counseling, which some obstetricians recommend. Although these procedures have been subject to some criticism, many view them to be beneficial.

In 1960, abortion, except to save the life of the mother, was illegal in every state in the United States. During that decade, while 8,000 therapeutic abortions were performed in the country, it is estimated that as many as 1 million illegal abortions were performed each year. Among women receiving illegal abortions, on average 350,000 suffered complications that resulted in hospitalization, and 500 to 1,000 died. In response, the National Association for the Repeal of Abortion Laws (NARAL) was founded in 1969; and efforts by it, other organizations involved in the Women's Health Movement, and women's advocates, led to the Supreme Court's hearing the case *Roe v. Wade* in 1973, their ruling on which legalized abortion.

In 1960, the birth control pill, initially approved only for treatment of severe menstrual disorders, was approved for contraceptive use, although many individual states had laws banning birth control altogether. Organizations, including those affiliated with the Women's Health Movement, had some success in changing such laws. In 1970, feminists challenged the safety of the birth control pill introduced a decade earlier in congressional hearings. Barbara Seamans' *The Doctor's Case Against the Pill* raised public awareness of what feminists argued were "gross neglect in testing the pill before it was prescribed for women, medical problems associated with it, and a lack on the part of pharmaceutical companies and the medical profession of reporting such problems" (Geary 1995, 28). Blood clots, stroke, heart attack, and depression were among the problems associated with the pill. In the 1980s, pills containing lower doses of hormones were introduced, again at the urging of those involved in the Women's Health Movement and other organizations seeking to improve women's healthcare. Eventually, as we note in our discussion of biomedicine, in the United States agencies at the federal level were established that were tasked with insuring that women's health and health issues received the study and funding appropriate to them.

Biomedical Research

Medicine came to recognize the potential benefits of biomedical research in the 1980s, and large studies into specific diseases and conditions, and clinical trials of new drugs and treatments (surgical and nonsurgical), began. Some studies and clinical trials included large numbers of participants and yielded concrete results. Their findings often had significant effects on medical treatments and guidelines in relation to specific diseases and conditions. And often these findings – for example, that for men, taking small doses of aspirin reduced the likelihood of a second heart attack; and correlations found in men between high blood pressure, smoking, weight, and other factors and cardiovascular disease. These and other studies and clinical trials led to increased knowledge, and better treatments and interventions for many medical conditions and diseases.

But feminists, as well as members of Congress and individuals associated with the National Institutes of Health (NIH), have pointed to three problems related to women's health and health issues in biomedical research in the 1980s and 1990s: 1) the almost exclusive reliance on male participants – that is, non-representative sampling – in research studies and clinical trials focusing on diseases and conditions that affect women as well as men; 2) the extrapolation of findings from studies and trials that involved male participants, to women, and the problems that resulted; and 3) the priority placed on diseases and conditions actually or assumed to be more common among men, such as cardiovascular disease; and the smaller amount of research into conditions affecting only or mostly women, such as breast cancer. As we will see, some changes relevant to these practices and assumptions began to come about in the 1990s. We begin with the problem of women's lack of representation in most studies and clinical trials during the 1980s and early 1990s, and the reasons given to explain it.

Londa Schiebinger cites many more such studies than we can list. We mention two that illustrate the issue: the 1986 Health Professionals' Follow-up Study of heart disease and coffee consumption in 45,589 men and no women, and the Physicians' Health Study of Aspirin and Cardiovascular Disease conducted in 1982 involving 22,071 male physicians and no women (Schiebinger 1999). Feminists note that a report by the Government Accounting Office in 1990 was highly critical of the NIH's support of studies in which men were the only or primary participants – particularly given that in 1986 the NIH had itself called for including more women in such studies. That these guidelines had been largely ignored was well known (Tavris 1992, 98).

What explains the over-representation of men in research studies and clinical trials? Different, but compatible, reasons have been cited. One was a widespread assumption that women's hormonal cycles are variables that potentially impact the results of studies and, thus, male participants are preferable. But, feminists argue, because females and males *do* vary physiologically, eliminating these variations means that the results of many experiments cannot be applied to females "in any simple or direct way" (Schiebinger 1999, 99).

As much to the point, feminists argue, researchers have assumed that men do *not* undergo hormonal cycles, something recent research calls into question. It is now widely, if not universally, accepted that men have monthly and seasonal hormonal cycles; and generally accepted that men experience age-specific hormonal changes during puberty, and between the ages of 40 and 55. Now one can note that there wasn't much by way of research into the possibility that men have hormonal cycles during the period in question, although there had been studies of the effects of low testosterone on men's sexuality, and into relationships between levels of testosterone and behavior and temperament. Given that androgens and some of their effects were identified early in the twentieth century, it is reasonable to ask why no studies were undertaken to determine if men, like women, had hormonal cycles until the late 1980s. Given the assumption that women's hormonal cycles would complicate research studies and clinical trials of drugs and treatments, why wouldn't researchers want to know if the same could occur in trials and studies involving men? Even if there are stronger and/or different variations in hormones cycles in women compared with men, feminists ask, wouldn't women's inclusion yield results that might prove important to knowledge about their health and response to treatments and drugs?

A different reason was given for excluding pregnant women and women of childbearing age from clinical drug trials that, at least on the surface, appears reasonable. NIH and biomedical researchers believed that having women in these groups participate in drug trials unnecessarily puts their fetuses and possibly pregnant women at risk. In the 1960s, the tragic consequences for fetuses whose mothers took Thalidomide (originally an over-the-counter sedative that some gynecologists reported as also relieving morning sickness and recommended to pregnant women), changed the way drugs were tested and approved under the auspices of the FDA. However, extrapolating results from research studies and clinical trials to pregnant women and those of childbearing age that only included male participants, feminists argue, *also* puts such women and their fetuses at risk.

A third reason given to explain the overrepresentation of men in research studies and clinical trials is that there are male populations – medical students, members of the military, and prisoners, for example – more easily "tapped" to participate in such studies, than others in the general population. But, of course, the demographics in terms of each of these populations had changed by the 1980s and 1990s, and would seem to undermine this argument.

The exclusion of women from biomedical research and the extrapolation of the findings of studies and clinical trials involving only or primarily men to them have had significant negative consequences for women, including severe reactions and even death. Again, we can only cite some examples that Schiebinger provides. Some drugs used to dissolve blood clots were found to be effective in treating men, but caused bleeding problems in many women. There is evidence that the effects of anti-depression drugs vary during a woman's menstrual cycle so that a constant dosage may be too high during some stages of it, and too low during others, something that a trial involving only men would not reveal. What have become standard drugs for treating high blood pressure in men, and successful in managing it, have been found to increase death rates in women. And the exclusion of women can lead to other problems. As Schiebinger concludes, "Not only are drugs developed for men potentially dangerous for women, drugs potentially beneficial to women may be eliminated in early testing because the test group does not include women" so that the benefits are unrecognized (ibid., 115).

Feminists also criticize the research priorities of biomedicine, noting an emphasis on "women's reproductive issues," rather than other health issues that affect women. (Schiebinger also points out that, despite the emphasis on women's reproductive issues, during the late 1980s the NIH had only three gynecologists and obstetricians on its regular staff.) Feminists have also criticized a lack of studies into how diseases affecting both men and women might present themselves differently in women than they do in men (cardio-vascular disease is often cited as one such example).

Fortunately, some important changes relevant to the problems feminists pointed to in biomedical research have occurred and are occurring. The NIH has now developed stricter guidelines to insure women's inclusion in research studies and clinical trials. It also established the NIH Office of Research on Women's Health, which has funded studies into previously understudied areas, such as sex/gender differences in auto-immune diseases and women's urological health. In 1991, the Women's Health Initiative began focusing on the causes, prevention, and treatment of leading killers of postmenopausal women, including cardiovascular disease, cancer, and osteoporosis. And efforts on the part of powerful members of Congress and women activists led to legislation in the 1990s that mandated women's health issues be granted the same priority as those of men. These issues are currently addressed through the NIH Office of Research on Women's Health (ORWH), which was established in response to concerns raised by members of Congress and women's health advocates in 1990. Under the auspices of ORWH, collaborative discussions and research involving scientific work-groups have led to several studies of how to include women of childbearing age and who are pregnant, as well as other women, so that research studies and clinical trials effectively and positively contribute to women's health.

Evolutionary Medicine

We turn now to a brief discussion of Evolutionary Medicine, which, as its name suggests, emphasizes the relevance of human evolution to human health and disease. Those who work in the discipline bring concepts and hypotheses of contemporary evolutionary theorizing to bear on medical issues, including natural and sexual selection; fitness; survival; reproductive success; what has come to be called the "arms race" between our species and infectious agents; human evolutionary history; and ultimate as well as proximate causes. The origins of Evolutionary Medicine are often credited to evolutionary biologist George C. Williams, and physician and evolutionary theorist, Randolph Nesse. In the early 1990s they co-authored articles and a book that called for an approach to medicine firmly grounded in evolution-ary theory (e.g., Nesse and Williams 1994; Williams and Nesse 1991). Evolu-tionary Medicine is of interest to feminists because of how it might change medical approaches to women's health issues (Zuk 1997) for reasons we briefly explore.

Although its focuses on disease and health parallel those of traditional medicine, because it emphasizes evolution, Evolutionary Medicine has led to new explanations of specific medical conditions and recommendations for their treatment. One oft-cited example is fever. Is it a medical condition to be treated using drugs that reduce it, as traditional medicine has viewed it? Or is it an adaptation that functions to speed up the body's immunological system to fight infection, as Evolutionary Medicine views it? Of course, there are dangers associated with very high fevers; but in cases in which the fever is not high enough to cause other problems, Evolutionary Medicine advises against the administration of fever-reducing drugs. As the editors of the textbook, *Evolutionary Medicine and Health: New Perspectives* (Trevathan et al.,

2008) point out, traditional medicine's approach to the issue reflects a focus on "proximate causes," and Evolutionary Medicine's argument for not treating non-life threatening fevers reflects a focus on "ultimate causes" – a focus, we have seen, central to much evolutionary theorizing (11).

We briefly consider several hypotheses about menstruation offered in Evolutionary Medicine and consider how they differ from the traditional medical approaches we have so far discussed. Although beyond the scope of our discussion, Evolutionary Medicine's approaches to issues that can arise during pregnancy often differ from traditional approaches (ibid., 21–24), as do approaches some in the field take to menopause (e.g., Stevert 2006).

An emphasis on evolution is reflected in *the question* Evolutionary Medicine asks about menstruation: "*Why* do women menstruate?" Biological anthropologist Lynnette Leidy Stevert addresses this question in an analysis focusing on whether women should take menstrual-suppressing oral contraceptives (MSOCs), rather than other kinds of birth control pills (Stevert 2008). On average, women who take MSOCs menstruate every three months, rather than monthly. Some women use MSOCs not only or even primarily to prevent pregnancy, but to suppress or regulate their menstrual cycles.

Many women might welcome fewer menstrual cycles, but there are disagreements in Evolutionary Medicine about whether this is potentially harmful. In brief, those who do not view the use of MSOCs as problematic point to the generally accepted hypotheses about ancestral women, according to which they would have had fewer menstrual cycles due to frequent pregnancies and longer periods of lactation than do women living in contemporary industrial societies. Based on studies of women in some contemporary hunter-gatherer groups, it is estimated that ancestral women experienced something like 94 menstrual cycles during their reproductive years, compared to the 400–500 such cycles typical of many contemporary women. In addition, some in the field have argued that the hundreds of menstrual cycles contemporary women experience may be pathological, "a *mismatch*" between our bodies and novel aspects of the environment, novel that is since the development of agriculture and especially since the Industrial Revolution (Nesse and Williams 1999, quoted in Stevert 2008, 182). More sedentary lifestyles and increased food security, they propose, resulted in more menstrual cycles. Some who don't view menstrual suppression as problematic also point out that during each menstrual cycle increased estrogen causes increased cell division in women's breasts, ovaries, and uteruses, and that cells that frequently divide are more likely to become malignant. Such arguments, Stevert notes, suggest that frequent menstruation isn't necessary to women's health, and may even be detrimental to it. From such perspectives, the effects of suppressed menstruation are likely not harmful.

On the other hand, several evolutionary explanations of menstruation Stevert considers, of which we mention three, complicate the issue. One hypothesis proposes that menstruation is an evolutionary by-product: it was not itself selected for, but came about due to the selection of a uterus that was not always ready for implantation; instead the uterine lining needed for implantation regrows every 28 days. Accordingly, "the functional significance of regression (menstruation) is metabolic economy," and the effects on women's health of taking MSOCs are neither beneficial nor harmful (ibid., 186). If menstruation is not an adaption and pregnancy is being avoided, "then it doesn't matter whether a woman menstruates every month or every 3 months" (ibid., 186).

Two other hypotheses take menstruation to be an adaptation. Marjorie Profit has proposed that menstruation was selected for to remove pathogens transported by sperm into women's lower reproductive tract (Profit 1993). Stevert noted that in 2008 Profit's hypothesis had not yet been substantiated (ibid., 187). But she also notes that if Profit is correct, suppressing menstruation is harmful. Another hypothesis is that menstruation is an adaptation for eliminating defective embryos and, thus, avoids maternal investment in them (Clarke 1994). If this hypothesis is correct, Stevert argues, the effects on women who use MSOCs to avoid pregnancy are neither beneficial nor harmful.

What is instructive for our purposes is that none of the approaches to or hypotheses about menstruation just summarized takes menstruation to be an "illness," as nineteenth-century medicine viewed it, or as representing "loss" or "failure," as twentieth-century textbooks imply. Rather, the researchers Stevert cites ask whether menstruation is an adaptation or a by-product of an adaptation that, in some way or other, either contributes to women's reproductive success or health, or if its current frequency represents "a mismatch" between women's bodies and contemporary environments. We have not, of course, considered every hypothesis about menstruation offered in Evolutionary Medicine. However, assuming an evolutionary viewpoint to ask "Why do women menstruate?", appears to be a more positive approach to that phenomenon than those that have been assumed by traditional medicine.

On the other hand, some hypotheses in Evolutionary Medicine assume Parental Investment Theory and other aspects of evolutionary theorizing feminists have criticized (for example, the assumption that contemporary hunger-gatherer groups provide insight into ancestral populations). On balance, however, one reason some feminists engage in or are interested in Evolutionary Medicine is that it may lead to more positive approaches to and better understandings of women's biology and health issues. Writing about the field in 1997, feminist biologist Marlene Zuk was cautiously optimistic that the differences in the questions asked by Evolutionary Medicine (the emphasis on "Why?" questions) would lead to more fruitful approaches to issues involving women's reproductive health (Zuk 1997).

Has Feminism Changed Medicine?

And if the answer to this question is "yes," to what extent, and what further changes are needed?

Obviously, the Women's Health Movement and activism on behalf of women's health have brought changes to aspects of medical understandings of and approaches to women's health. As we have seen, they did so, in part, by educating women about their reproductive organs and processes, by challenging medical views and treatments of things like pregnancy and childbirth, and by encouraging women to question and, if necessary, challenge the advice and treatments offered them by obstetricians and gynecologists. We have considered only some of the changes feminist activism resulted in, most involving the ways in which the conditions in which labor and childbirth occurred changed for many women, how birth control and abortion became (at least, in principle) available to many women, and how feminists and others brought attention to the medical problems associated with the first birth control pill. (They did the same, by the way, in terms of the first IUD.)

Schiebinger argues, however, that changes at the federal level of government brought about by seasoned and effective politicians, have done as much to change medicine in ways conducive to women's health and in keeping with feminists' goals, as have feminist activism and the increase in the number of female physicians. She argues that many other factors and players had at least as large a role in bringing about changes in terms of medicine and women. She points out, for example, that the influence and success of the Women's Health Movement were due in part to broader social change that led to "The Equal Opportunity Act" of 1972, and "The Equal Opportunity in Science and Engineering Act" of 1980 (Schiebinger 1999, 122). The latter, she points out, specified that the NIH and the National Science Foundation increase the number of those in underrepresented groups, including women and minorities, in medicine and science. As another example, she cites congressional interventions, including the creation of and work done by the Congressional Caucus on Women's Issues, of which many politically influential women were members, as being of significant help in bringing about the reforms feminists called for. It seems reasonable to conclude that many factors, including social and political climate as well as feminist efforts, contributed to changes in medicine in terms of women's health and health issues.

To the question, "Is there more that needs to be done?" it is likely that most feminists would say "yes." There are women in Western industrialized countries who, because of socioeconomic status and geographical location, have not benefitted nearly as much as others. There are continuing efforts to legislate women's reproductive options, including to limit birth control and access to (if not a total ban on) abortion that need to be addressed. There are illnesses and diseases afflicting women about which much more research is needed. There are, feminists and others argue, more hysterectomies and Caesarian sections than the evidence indicates are necessary. There is also a need to continue to challenge medical assumptions about processes unique to women that have a direct and consequential impact on women's health. Two recent developments are telling in this regard: what some feminists argue is basically a regression in terms of how menstruation and menopause are viewed and treated, even if many medical practitioners understand the regimens of treatment they now recommend as substantively contributing to women's health.

By the 1970s, many gynecologists and women came to accept that a syndrome, Premenstrual Syndrome (PMS), affects many women, and should be medically treated. But feminists and many gynecologists find the definition of the syndrome, which consists of a long list of symptoms, and its attribution to many women, questionable. To be sure, there are women for whom the days before and during menstruation are characterized by more serious and numerous symptoms than most other women experience. But when some researchers and medical practitioners attribute the syndrome –

recall that "a syndrome" is a group of symptoms that consistently occur together – to any woman experiencing even *one* such symptom and, on that basis, estimate that as many as 85 percent of women suffer from PMS. We seemed to have regressed to the view of earlier eras that menstruation is an illness. At the very least, feminists and others argue, more research into this "syndrome" and how many women it affects, is called for.

Similarly, the emergence of Hormonal Replacement Therapy (HRT) to treat menopausal and postmenopausal women assumes the definition of menopause as "estrogen-deficiency" that warrants medical intervention, rather than – for many if not most women – a natural stage of life. Once it was so defined, many physicians and pharmaceutical companies advocated estrogen replacement for menopausal women. But in 2002, studies about the effects of HRT on women led to questions and debates about its use. Even though treatments have improved, medical organizations and physicians consistently list its "pros" and "cons," and recommend against its use for women at higher risk of breast or ovarian cancers. There are also women and physicians who now choose alternative ways to treat symptoms some women experience at menopause.

So, it seems reasonable to conclude that, given episodes such as those just mentioned, given as well the state of knowledge and of treatments for diseases that disproportionately affect women – including breast cancer and osteoporosis – there continues to be a need for more research into, and more changes in medical approaches to and treatments of, health issues affecting women.

7   Neurobiology

Are There "Male Brains" and "Female Brains"?

Since the mid-twentieth century, research devoted to identifying differences among male and female brains, including men's and women's brains, has been undertaken in a variety of sciences that study the brain and nervous system (commonly referred to as "the neurosciences"). These include neuroendocrinology, behavioral endocrinology, empirical psychology, and medical psychology. In this chapter, we focus on what has come to be called "Brain Organization Research" (hereafter, "BOR") and several "brain organizer hypotheses." These hypotheses propose that differences in the levels of testosterone to which males and females are exposed during development lead to differences in how male and female brains are "organized," and that these differences explain what those advocating the hypotheses take to be well established sex/gender differences beyond the anatomical differences related to reproduction. Individually, the hypotheses we consider claim to explain purported sex/gender differences in mathematical abilities, behavior and temperament, and "male" and "female" heterosexuality.

BOR is not the only research program in the neurosciences investigating whether and how men's and women's brains are different. Nor is it the only program in neuroscience that feminists critically engage. We focus on brain organizer hypotheses because they have been influential, widely publicized, and are well respected; and because they are accessible to non-specialists. We emphasize aspects of the reasoning that lead to and inform them, and feminist critiques of that reasoning.

Some readers might find the idea that there are "female brains" and "male brains" curious. After all, we don't think that all physical features that, on average, differ along the lines of sex, are themselves "male" or "female." For example, on average, women's hearts are smaller than men's. We might find it reasonable to say that this constitutes an average difference between women and men, but it does not follow from this that there are "female hearts" and "male hearts." It is also possible that some scientists who use the phrases "male brains" and "female brains" are speaking metaphorically. But this is not true of all scientists who use these phrases, and it is certainly not true of those whose hypotheses we consider.

We will find that some scientists, including but not limited to feminist scientists, view investigations of differences in men's and women's brains to be motivated by social and political interests (e.g., Bleier 1984; Kinsbourne 1980). But we will also find that there are general hypotheses accepted in several research programs that have been understood to provide a scientific rationale for such investigations. Three are particularly relevant to the brain organizer hypotheses we consider.

Neuro- and reproductive endocrinology study relationships between hormones, brains, and the nervous system. Because males and females are exposed to different amounts of androgens and estrogens prenatally (and there is evidence that this also occurs postnatally), studying how such differential exposure results in anatomical differences related to sexual reproduction can provide a useful "baseline" from which to study the effects of hormones more generally. That differential exposure to sex hormones during fetal development contributes to such anatomical differences is well established as we discussed in Chapter 5. In addition, anatomical abnormalities that result from excessive or deficient prenatal exposure to "sex-appropriate" hormones also provide evidence of their effects (Longino 1990).

Behavioral endocrinology studies relationships between behavior, and the phenomena on which neuro- and reproductive endocrinology focus. Given the causal effects of differential exposure to sex hormones on anatomical differences in males and females, some in these disciplines take it to be highly likely that this differential exposure also has an effect on brains and such effects cause sex and sex/gender differences in behavior. If one assumes that sex/gender differences in behavior and temperament are well established, then research into how prenatal hormone levels might result in such differences would seem warranted.

There is also substantial research focusing on laboratory animals, particularly rodents, that investigates the effects of sex hormones on brains, behavior, temperament, and "skills," including sex differences in the species studied. Unlike studies of human fetuses and adults, the laboratory studies of nonhuman animals often involve experiments that manipulate sex hormone levels to identify their effects. Those who have developed the hypotheses we consider about differences in the organization of men's and women's brains, and the effects of these purported differences in terms of phenomena such as behavior, often cite the findings of studies of laboratory animals as evidence that the sex/gender differences they propose to explain do exist.

Evolutionary theory is also taken to provide a general rationale for investigating differences between men's and women's brains. We have seen that many evolutionary theorists believe that sex/gender differences in temperament, abilities, behavior, and heterosexual sexuality are highly likely because these theorists assume that our female and male ancestors faced different adaptive problems. And, as we learned in earlier chapters, there are evolutionary theorists who view such differences to have been "selected for" because they were conducive to reproductive success. Some maintain that the mating and parenting strategies of males and females, including those of men and women, conflict. Others take sex/gender differences in terms of temperament, behavior, and the like, to be "complementary." That such differences are rooted in the brain is viewed as at least a plausible, if not a likely explanation of them.

So, although there may well be sociopolitical interest in establishing that men's and women's brains are different and that this results in differences in behavior or mathematical ability, the general hypotheses just summarized are understood by many scientists to warrant investigations into whether and why women's and men's brains are different. From a contextualist perspective, the fact that social and scientific contexts may both contribute to interest in this issue is not surprising. But we will also consider critiques offered by feminists that challenge the general scientific hypotheses just mentioned.

Some qualifications are in order. The hypotheses we consider maintain that the differences between women's and men's brains and their effects are significant. But although reports of them have been widely disseminated, including in publications written for the lay public, there isn't consensus among neuroscientists that there are such differences or, if there are, that they are significant. Some reject the idea outright. Still others maintain that although there are such differences, for example, in mathematical ability, the differences between the sex/genders are less than the differences found to exist *within* each group. If this is the case, then knowing only an individual's sex/gender does not allow predictions concerning that individual's mathematical abilities.

Second, in the past decade, feminist neuroscientists and cognitive scientists have studied and offered sustained critiques of hypotheses, technologies, and methodologies more recent than those we consider. Both the research in question and the critiques feminists offer of it are important (e.g., Bluhm et al. 2012; Jordan-Young 2010). But they are more technical and complex than we can do justice to here.

Philosophical Issues

The philosophical issues we consider are related to aspects of the reasoning that underlies and informs brain organizer hypotheses, and to the critiques of and alternatives to that reasoning that feminists offer. And many of these philosophical issues are ones we discussed in earlier chapters.

One, discussed in Chapter 5 in relation to the 1980s account of sex determination, involves "theoretical (or cognitive) virtues" that studies of scientific practice indicate scientists value. (Some philosophers of science, including feminists, use the phrase "cognitive values" [e.g., Longino 1996] to refer to what we are calling theoretical or cognitive virtues.) Brain organizer hypotheses offer explanations that are linear and hierarchical. Recall that an explanation is "linear" if it proposes unidirectional causal relationships between entities and/or processes, and an explanation is "hierarchical" if it proposes that one or at most a few entities or processes causally determine intermediate and ultimate effects. Explanations of this sort have long been taken to be characteristic of "mature science." That is, their form has been viewed to be a theoretical virtue. Such explanations exhibit "simplicity," a feature that studies of scientific practice indicate is also viewed as a theoretical virtue by many scientists. We consider feminists' arguments that the linear, hierarchical form of brain organizer hypotheses oversimplifies complex biological relationships and fails to take relevant social and environmental causal factors into account. With regard to the latter, feminists are criticizing the biological determinism assumed in BOR. Most feminists *do not argue* that linear, hierarchical explanations are, in principle, inappropriate when the causal relationships in question are biological. There are explanations of this sort for which there is sufficient evidence. In rats, for example, differences in prenatal exposure to androgens and estrogens affect the development of sex differences in the region of the brain below the cortex that is related to estrus cyclicity (e.g., Longino 1990). However, feminists do criticize brain organizer hypotheses that are of this form for the reasons cited above and discussed in detail in forthcoming sections.

Feminists critical of brain organizer hypotheses also appeal to the underdetermination thesis we discussed in Chapters 1 and 2, and specifically to Hempel's and others' thesis that unrecognized and/or unexamined background assumptions are often features of scientific theorizing. In their arguments that brain organizer hypotheses are underdetermined by available evidence, feminists often point to the presence of such background assumptions in BOR experiments and hypotheses. What feminists add to Hempel's argument and those of others who recognize the role of background assumptions in scientific reasoning is that these can include unquestioned and unwarranted assumptions about sex/gender, including gender stereotypes, gendered metaphors, and evaluatively thick concepts. If they are correct, then background assumptions are not, as many philosophers have assumed, limited to "purely scientific" assumptions. Again, from the perspective of feminist versions of contextualism, that such background assumptions can inform research into sex/gender differences is not surprising. Nor as we noted in Chapter 4, contrary to more traditional accounts of science (e.g., Haack 1993), do feminists take it to entail that science so informed can be written off as "bad science." Underdetermination, and how it leads to the role of background assumptions in scientific reasoning, have been mentioned in relation to research we discussed in earlier chapters and are relevant to research we consider in subsequent chapters.

Although it is beyond the scope of our discussion, there is a relationship between underdetermination theses (there are several versions of that thesis) and the recognition that theoretical virtues have a role in scientific practice. Hypotheses that exhibit such virtues are taken to be more plausible than competing hypotheses that lack them (e.g., Duhem 1954).

A third philosophical issue involves the difference between what have come to be called "quasi experiments" and "true experiments." Briefly put and further explicated in what follows, studies of human subjects are classified as "quasi experiments" because, for both ethical and practical reasons, such studies cannot control for many variables that might be causally relevant to the phenomenon being studied. Another reason such studies are

"quasi" rather than "true" experiments is that it is not acceptable to manipulate or alter biological traits of human subjects. In contrast, true experiments, such as those involving laboratory animals, do control for many of
the variables that may be relevant to the research questions being pursued.
And, as earlier noted, studies of laboratory animals often involve manipulation of their biology, something that studies of human subjects, for ethical
reasons, cannot undertake. We consider feminists' arguments that the evidence that studies of human subjects can provide, evidence to which brain
organizer hypotheses appeal, is insufficient to establish that the hypotheses
in question are warranted because they must rely on "quasi experiments."

In our discussion of primatology, we introduced the distinction many
draw between proximate and ultimate causes, and this is the fourth philosophical issue we consider. As we have noted, brain organizer hypotheses
propose that differential exposure to sex hormones during development
results in sex/gender differences in brain organization, and these in turn
lead to differences in women's and men's behavior, or temperament, or
cognitive abilities, or (heterosexual) sexuality. We will see that feminists
point to other causal factors, including gender expectations and socializa-
tion, as significantly contributing to such sex/gender differences in cases in
which they do exist. Whether they argue that sociological factors are the
ultimate cause, or one of several proximate causes, most feminists challenge
the biological determinism assumed by brain organizer hypotheses – the
assumption that biology is the "ultimate" cause of the sex/gender differences
in question. We also consider how, as earlier noted, some evolutionary
theorists view natural or sexual selection as the ultimate cause of sex/gender
differences in abilities, behavior, temperament, and other characteristics,
and how some researchers in BOR also make this assumption.

We turn now to several specific brain organizer hypotheses, and the
criticisms of and alternatives to them feminist scientists and philosophers
have offered. The first proposes to explain why boys and men exhibit mathematical abilities that are superior to those of girls and women.

Sex/Gender Differences in Mathematical Abilities

In 1982, neuroendocrinologists Norman Geschwind and Peter Behan proposed that the higher level of exposure of male fetuses to testosterone causes
right hemisphere lateralization in boys and men. Building on studies

undertaken in empirical psychology and reproductive endocrinology that reported sex and sex/gender differences in lateralization in rodent and human brains, they proposed that right hemisphere lateralization explains the "superior performance" of males of various species in "spatial contexts" (Geschwind and Behan 1982). And in 1984, Geschwind and Behan further proposed that right hemisphere lateralization in boys and men also explains their superior "mathematical ability" relative to girls and women because such ability is related to spatial reasoning (Geschwind and Behan 1984). Here some background about hemispheric lateralization, and how hypotheses about sex and/or sex/gender differences in lateralization were of interest in empirical psychology beginning in the 1960s, will be helpful.

Our brains contain two hemispheres (a "right hemisphere" and a "left hemisphere") and there is evidence that each is somewhat more involved in some brain functions than is the other (although we will see that some criticize what they take to be an overemphasis on "hemispheric specialization"). The right hemisphere is generally associated with spatial abilities, the left with verbal abilities. Some of the evidence supporting the idea that each hemisphere is more implicated in some specific functions than is the other is provided by people who suffer damage to one hemisphere. It is well known, for example, that people who suffer damage to the right hemisphere lose some degree of spatial ability, at least for a period of time. If the damage is severe, the effects include an inability to see things to their left (often including the left side of their own bodies, control over which is also partially or completely lost) – phenomena referred to in the medical literature as "left neglect" – and an inability to recognize faces or to use clocks, as they perceive the features of these and other objects as "jumbled." Some effects of severe damage to the left hemisphere are also well known. They include an inability to control the right side of the body, to "see" objects to one's right including the right side of one's body (phenomena referred to as "right neglect"), and loss to some degree or other of language abilities.

In some cases, the effects of damage to a hemisphere are temporary, apparently because other parts of the brain, including the other hemisphere and the *corpus callosum* (a massive set of nerve fibers that connect the two hemispheres) eventually "take over" functions associated with the damaged hemisphere. That this occurs suggests that although lateralization may be characteristic of human brains, it is sometimes overcome or compensated for by other parts of the brain.

In their 1982 article, Geschwind and Behan appealed to studies undertaken in several sciences as support for their hypothesis that boys and men are right hemisphere lateralized. There were apparent correlations between migraines, immune system disorders, learning disabilities, and left-handedness. And there were studies that found that in each case, these are more common in boys and men than in girls and women. At the time, many thought left-handedness was caused by right hemisphere lateralization, although today there is evidence that specific genes also have a causal role in handedness.

Geschwind and Behan's 1982 hypothesis was that prenatal testosterone slows the development of the left hemisphere resulting in right hemisphere lateralization in males. They emphasized the results of two studies as providing evidence of this. In a study of rat brains, Diamond et al. reported that two areas of the cortex of male rat brains are 3 percent thicker on the right side than they are on the left, and that such asymmetry is not found in female rats. They also reported that they managed to bring about asymmetry in female rat brains by removing ovaries, and they could prevent the asymmetry in male rat brains by castrating males (Diamond et al. 1981). Appealing to then current hypotheses in empirical psychology – that right hemisphere lateralization increases visuospatial abilities – Diamond et al. proposed that such lateralization is characteristic of male rats and explains why they more are more successful in negotiating mazes than are female rats.

The hypotheses in empirical psychology to which Diamond et al. appealed were proposed beginning in the late 1960s. They maintained that boys and men utilize the right hemisphere when engaging in mathematics, and that this explains their superior performance in the mathematical tasks included in standardized tests. Interestingly, two of the hypotheses frequently cited at the time about lateralization in girls and women contradicted one another, although both claimed to explain girls' and women's poorer mathematical performance on standardized tests. One proposed that girls and women's brains were *less* lateralized than those of boys and men, which resulted in girls and women using both verbal abilities (that is, their left hemisphere) and visuospatial abilities (that is, their right hemisphere) when they engaged in mathematics (Levy-Agresti and Sperry 1968). The other proposed that girls' and women's brains were *more* lateralized, with the left hemisphere being dominant, and that this caused girls and women to use verbal abilities when engaging in mathematics (Buffery and Gray 1972). These differences notwithstanding, both were proposed to explain what were taken to be inferior mathematical abilities among girls and women compared with those of boys and men.

Geschwind and Behan also appealed to studies of human fetal brains undertaken by Chi et al. to support their extrapolation of Diamond et al.'s hypothesis and research findings to humans, and to support their own hypothesis that testosterone slows the development of the left hemisphere. Chi et al. reported that two convolutions of the right hemisphere of human fetal brains develop several weeks earlier than do corresponding convolutions of the left hemisphere (Chi et al. 1977). But, as we later discuss, they reported finding *no* sex differences in the human fetal brains they studied.

In extending the effects of right hemisphere lateralization to mathematical abilities in 1984, Geschwind and Behan appealed to a study undertaken by empirical psychologists Camilla Benbow and Julian Stanley reporting "a marked excess of males among mathematically gifted children" as additional evidence that the higher level of exposure to testosterone that male fetuses experience results in right hemisphere lateralization in boys and men, and that this explains their superior mathematical abilities relative to girls and women (Benbow and Stanley 1983, cited in Geschwind and Behan 1984). They again cited the then current (but contradictory) hypotheses in empirical psychology that women's brains were differently lateralized from men's that we noted above. Interestingly, Benbow' s and Stanley's 1983 article appealed to the 1982 article by Geschwind and Behan with which we began.

As the foregoing indicates, Geschwind and Behan's hypothesis that right hemisphere lateralization causes superior mathematical abilities among boys and men did not emerge in a scientific vacuum. At the time, there was substantial interest in studying sex and sex/gender differences in brains and in mathematical ability. As we have noted, evidence that differences in prenatal exposure to androgens and estrogens caused anatomical differences was taken to suggest that they have effects on brain function as well. Research in disciplines closely allied to behavioral endocrinology was also viewed as providing evidence that males' and females' brains are different. Seventeen years prior to Geschwind and Behan's 1984 hypothesis, studies in reproductive endocrinology reported that androgens block the cyclical response of hypothalamic neurons that regulate pituitary functions related to estrous cyclicity in female rats. Some years later, researchers in the discipline reported morphological sex differences in the brains of some bird species in areas related to the ability of males to sing. On the basis of such findings, neurobiologists Gorski et al. (1978) maintained that "the concept of the sexual differentiation of brain function is now well established" and called for further investigations of it.

Feminist Critiques and Contributions

Geschwind and Behan's hypotheses that right hemisphere lateralization results in superior mathematical abilities in boys and men is obviously a linear, hierarchical explanation. Fully specified it proposes that:

> In male fetuses, the Y chromosome secretes H-Y antigens → H-Y antigens contribute to the development of male internal gonads → Male internal gonads secrete testosterone → Testosterone causes right hemisphere lateralization by slowing the development of the left hemisphere → Such lateralization results in superior visuospatial abilities → Superior visuospatial abilities result in superior mathematical abilities.

Although they did not cite testosterone as causing right hemisphere dominance, the earlier hypotheses of empirical psychologists about sex/gender differences in lateralization to which Geschwind and Behan appealed, were also linear and hierarchical as they proposed that sex/gender differences in lateralization cause differences in visuospatial reasoning and mathematical ability. So, too, was Diamond et al.'s hypothesis about sex differences in rodent brains, according to which prenatal exposure to testosterone causes thickness in areas of the right side of the cortex in male rat brains and such thickness is causally related to right hemisphere lateralization.

Although we will see that this is not the only issue feminists raise about brain organizer hypotheses, they have devoted a lot of attention to the emphasis placed on linear and hierarchical causal relationships. Feminist biologists argued that the causal relationships the hypotheses we have considered proposed, were not established. Of Geschwind and Behan's hypothesis, they argued that no causal mechanism had been identified to explain a relationship between testosterone and slower development of the left hemisphere. They also argued that although Diamond et al.'s research might establish more thickness in the right side of the cortex in developing male rodent brains than is characteristic of female rodent brains, it did not establish a causal relationship between that thickness and right hemisphere lateralization (e.g., Bleier 1984).

And feminists pointed out that Geschwind and Behan misstated the results of Chi et al.'s study of human fetal brains they cited as supporting their hypothesis that testosterone causes right hemisphere lateralization (Bleier 1984). Chi et al. did report that convolutions in the right hemisphere appear earlier than in the left in human fetal brains. What Geschwind and Behan did not state in appealing to the study, is that it reported the differential development in *both* male and female brains. Indeed, Chi et al. stated that the scientists involved in the studies in question "could recognize no difference between male and female brains of the same gestational age" in the 507 human fetal brains of 10–44 weeks' gestation they studied and measured (Chi et al. 1977, 92).

In their general challenges to the linear, hierarchical explanations we have considered, feminists criticized the simple and uni-directional causal account of *human* brain development, citing experimental results discussed in Chapter 5 that indicate complex and often non-linear interactions between cells, and between cells and the maternal environment, during every stage of human fetal development. In a related line of argument, they challenged the extrapolation of results of studies of rodent brains to human brains. Research in neurobiology and neurophysiology, they pointed out, indicates that human brains develop over a much longer period of time than do the brains of simpler species such as rodents, that much of this development occurs postnatally (specifically during the first two years of life), and that this development *requires* specific and intense interpersonal stimuli (e.g., Bleier 1984). They also cited the much greater complexity of human brains compared to those of simpler mammals, and their greater plasticity. Over the course of our lifetimes, our experiences, including learning, result in significant changes in neural connections and pathways. For these several reasons, feminists argued that the assumptions about human brain development made by Geschwind and Behan are unwarranted, as is extrapolating from the development of rodent brains to draw conclusions about how human brains develop (e.g., Jordan-Young 2010). In these lines of argument, feminists criticized the appropriateness of and warrant for the biological determinism that linear, hierarchical models of the relevant biological processes assume.

Feminists' critiques of the reliance on studies of brain development and behavior in rats to hypothesize about human brains and behavior raised additional issues. One is the presumed relationship between male rats' "maze-negotiating abilities," reported as superior to those of female rats, and human mathematical abilities. In more general terms, feminists argued that the notion of "visuospatial abilities" was not operationalized in ways that warranted its extrapolation across species. Feminist biologists also argued that no clear relationship had been articulated, let alone established, between "visuospatial abilities" and mathematical ability (e.g., Bleier 1984).

In addition, feminists noted that hypotheses in empirical psychology proposing sex/gender differences in lateralization to which Diamond et al., and Geschwind and Behan, appealed, were controversial within that field. As earlier noted, two widely cited hypotheses contradicted each other about precisely how girls and women differ from boys and men in terms of lateralization. And in a review article of research seeking to identify sex/gender differences in lateralization, Marcel Kinsbourne, well known for his research on hemispheric lateralization, concluded that the evidence for the sex/gender differences being claimed "fails to convince on logical, methodological, and empirical grounds." He proposed that the reason "reputable investigators" ignored what he took to be quite obvious problems was "a determination to discover that men and women 'really' are different ... given the growing momentum of the feminist movement" (Kinsbourne 1980, 242). Whether Kinsbourne' s explanation of why sex/gender differences in lateralization were pursued in empirical psychology is correct, the issue of borrowing hypotheses and reported research findings, which characterized much of the reasoning we have considered, is important. Such borrowing is not, of course, a mark of "bad" or flawed science; indeed, it can lead to significant advances (Darwin provides a clear example of this). What is problematic, feminists argued, are cases in which the fact that the hypotheses being cited are controversial, and in one case, even contradictory, is not noted by those appealing to them. Misstating the results of a study to which one appeals, as happened in Geschwind and Behan's appeal to Chi et al.'s findings, is also problematic.

Finally, feminists questioned the rationale of looking for a biological foundation for the sex/gender differences in mathematical abilities that Geschwind and Behan's hypothesis, and those in empirical psychology concerning lateralization to which they appealed, were taken to explain.

Feminists cited studies indicating more variance in "measures" of mathematical ability among members of each sex/gender, than between them (e.g., Fausto-Sterling 1985). They also pointed to changes in social expectations and education policies in the late 1970s and 1980s that resulted in a closing of the gap between girls and boys in mathematical performance on standardized tests (a gap sex-differentiated brains were initially proposed to explain), and noted that the sex/gender differences in math scores varied by test. And they challenged the equating of "mathematical ability" with performance on standardized tests. Noting that girls and women typically receive as high grades in math classes as do boys and men, feminists asked why grades were not taken to be indicators of mathematical ability? Feminists also argued that "the tasks" emphasized in experiments to determine if there are sex/gender differences in such ability are "selective," focusing on those involving mental spatial rotation rather than computation. In general, girls and women have been less successful in tasks involving mental spatial rotation but no less successful in performing computational tasks. It seemed clear to many feminists that gender stereotypes informed the assumptions about sex/gender differences in mathematical abilities. In more general terms, in the several lines of argument considered in this section, feminists challenged the empirical warrant for Geschwind and Behan's hypothesis that prenatal hormones cause superior mathematical abilities among boys and men (Nelson 1993; Jordan-Young and Rumian 2012).

Some of the assumptions that characterized the research from the 1960s through the 1980s we have discussed have been challenged and some even abandoned. For example, some neuroscientists report that men's brains are more lateralized when engaging in tasks involving language than are women's, although studies using fMRI imaging to investigate this hypothesis often yield conflicting or inconclusive results (e.g., Kansaku and Kitizawa 2002). And some attempts to replicate the findings reported in earlier studies of clear sex differences in brain *function* in laboratory animals such as hamsters – including in lateralization –have been unsuccessful (e.g., Schum and Wynne-Edwards 2005). Nor have hypotheses that males of many non-human species consistently exhibit superior spatial abilities been confirmed (Costanzo et al. 2009). And Geschwind and Behan reported that some of the correlations they proposed between right hemisphere lateralization and conditions more common in boys and men cited in their 1982 article were not confirmed in subsequent studies (Geschwind and Behan 1984).

In addition, an experiment undertaken at UCLA in 2007 suggests that women's poorer performance on tasks involving mental spatial rotation – "the most reliably observed sex/gender difference in cognitive skill ... that consistently favors males" – may be a function of experiences that often differ by sex/gender (Jordan-Young and Rumian 2012, 110). Researchers reported that this sex/gender difference "was virtually eliminated" when young women were exposed to about 10 hours of training in video games. Their scores improved much more than did those of men who underwent the same training and much more than did those of students who did not undergo the video game training (ibid., citing Feng et al. 2007). Those who performed the experiment hypothesize that its results may reflect relative inexperience among young women as compared with young men in playing video games. It remains to be seen if the results will be replicated in further studies.

We need to note, however, that sex/gender differences in performance on mathematical problems on standardized tests *predate* the existence of video games. Were there other differences in boys' and girls' experiences before there were video games that could explain the higher performance of males on standardized tests? Speaking from firsthand experience as a woman raised in the 1950s and 1960s, I suggest there is. At that time, boys and girls were expected to, and typically did, engage in very different activities. I was never encouraged to play "catch" or engage in other sports; and my toys, unlike my brothers, did not include precursors to today's Lego toys. My experiences as a girl growing up in that era were quite typical. Given the brain's plasticity, it seems plausible that some of the activities and toys to which boys were exposed more often than girls enhanced their visuospatial abilities. But I should also add that, in comparison to my male counterparts in K–12 and university education, I did as well as most and better than many in terms of mathematical performance on standardized tests. I suggest that the explanation is that my father, a mathematician, believed in my mathematical abilities and pushed me to succeed. That is to say, my success was no doubt due in large measure to personal social factors that many other girls of my generation did not experience.

In summary, the feminist critiques and contributions discussed in this section reflect an argument that an empirically adequate explanation of sex/gender differences in mathematical performance on standardized tests and/or in mathematical ability in general, would take into account the many relevant factors feminists cite, rather than ignoring them, and proposing linear, hormonal explanations. Feminists argue that unquestioned and/or unexamined assumptions about sex/gender mediate the distance between these hypotheses and available data.

But although some assumptions about sex/gender differences in brain lateralization and their effects on mathematic abilities have changed since Geschwind and Behan's 1984 paper, Geschwind and his collaborators have continued to develop what has come to be called "Geschwind's Theory" (which maintains that there are important sex/gender differences caused by differences in lateralization) and their research continues to be influential in some disciplines. It seems likely that there will be further developments of it, and further critiques of it, in the future.

Sex/Gender Differences in Behavior and Temperament

We turn now to brain organizer hypotheses that propose that the differential level of exposure of human male fetuses and female fetuses to prenatal sex hormones, and some suggest to postnatal sex hormones, cause sex/gender differences in brains that result in differences in specific behaviors and kinds of temperament. Because those proposing the hypotheses we consider typically focus just on pre- and postnatal exposure to testosterone during male development, rather than on "higher levels" of exposure to testosterone than common for females, we do so as well, but remind readers that male and female fetuses are exposed to both androgens (including testosterone) and estrogens, albeit to different amounts.

The hypotheses we consider are representative of many that have been proposed since the 1970s and 1980s. Like the hypotheses discussed in the previous section, hypotheses proposing causal relations between prenatal hormones, brain differences, and sex/gender differences in behavior and/or temperament developed as studies of laboratory rodents reported finding sex differences in behavior and temperament linked to differential exposure of males and females to sex hormones. These studies, as we noted earlier, often involved experiments that manipulated the levels of exposure of rodents to sex hormones. They reported increasing the exposure of male rodents to estrogens resulted in their exhibiting behaviors typical of females (such as lordosis), and increasing the exposure of female rodents to testosterone resulted in their exhibiting behaviors typical of males (such as mounting).

Such studies and results have been appealed to by those advocating brain organizer hypotheses proposing sex/gender differences in behavior and temperament. In addition, many studies undertaken since the 1960s of girls and boys, and men and women, report sex/gender differences in behavior and/or temperament.

Of course, not all forms of behavior or kinds of temperament have been assumed or argued to be causally related to pre- or postnatal exposure to sex hormones. The behaviors and kinds of temperament at issue are those believed to differ between males and females, including men and women. One common hypothesis is that pre- and postnatal exposure to testosterone is causally related to "aggression" and "aggressivity," behavioral and temperamental traits generally assumed to be more common among males, including boys and men. The conclusion endocrinologists F.H. Bronson and C. Desjardins drew, based on their study of laboratory mice in which males were observed to be far more aggressive than females, is representative.

> We may expect both the organizing and adult modulating roles of testosterone to be important in any species in which there exists a reasonable sexual difference in aggressiveness in favor of the male. (Bronson and Desjardins 1976, 101)

In terms of humans, psychologists Eleanor Maccoby and Carol Jacklin, although described by feminist philosopher Helen E. Longino as "nonpatriarchal scholars," agreed that, in general, men exhibit more aggressive behavior and that this is likely to be causally related to, although perhaps not solely determined by, pre-and postnatal exposure to testosterone (Longino 1990, 116 citing Maccoby and Jacklin, 1974, 244).

Some scientists have also proposed that "male aggression" explains "male dominance." As we have seen, this hypothesis was advanced by some Human Sociobiologists, but it also enjoyed broader acceptance. In 1973, anthropologist Steven Goldberg took the findings of studies of laboratory rodents, as well as brain organizer hypotheses linking testosterone to aggression among men, as proof that the emerging feminist movement could not succeed.

> Human biology precludes the possibility of a human social system whose authority structure is not dominated by males, and in which male aggression is not manifested in dominance and attainment of positions of status and power. (Goldberg 1973, 78)

We earlier noted that psychologists Maccoby and Jacklin agreed that, on average, boys and men exhibit higher levels of "aggressive behavior" and that there is likely a causal relationship between it and pre- and postnatal exposure to testosterone. They were, however, less convinced that, in humans, there is a relationship between "aggression" and "leadership" or "dominance" (Maccoby and Jacklin 1974, 245).

In 1981, medical psychologists Anke Ehrhardt and Heino Meyer-Bahlburg published a paper based on studies they undertook of girls and women with a condition known as Congenital Adrenal Hyperplasia (or CAH). These girls and women had been exposed to much higher than normal levels of androgens, including testosterone and dihydrotestosterone, during fetal development, either because their adrenal glands produced excess amounts of androgens and failed to produce cortisone, or because their mothers were treated with a high dosage of progestin during pregnancy to prevent miscarriage. Ehrhardt's and Myer-Bahlburg's studies led them to propose that girls and women with CAH exhibit behavioral and temperamental traits more typical of boys and men than of girls and women.

The reasoning involved in this case is different from that characterizing the brain organizer hypotheses we previously considered. Rather than starting from studies of behaviors exhibited by males and females to hypothesize that the sex or sex/gender differences observed are caused by differences in exposure to pre- and postnatal sex hormones, Ehrhardt and Meyer-Bahlburg studied the behavior and temperament exhibited by girls and women *known* to have been exposed to higher than normal levels of androgens during fetal development. Before discussing their research, we briefly summarize the anatomical effects of CAH, and the surgical and therapeutic interventions intended to at least minimize them. This is because feminists who are critical of the linear, hierarchical explanation Ehrhardt and Meyer-Bahlburg propose argue that the psychological effects of CAH and medical interventions are equally plausible explanations for the behaviors and temperament Ehrhardt and Meyer-Bahlburg observed.

Girls with CAH are born with clitorises larger than those of other girls and women (sometimes large enough that they can be mistaken for penises) and incomplete fusion of the labia. And unless they receive estrogen treatments, they exhibit degrees of "virilization" when they reach puberty. Most undergo at least two corrective surgeries, one in infancy and another at puberty. At the time Ehrhardt and Meyer-Bahlburg undertook their studies, the surgeries typically involved complete or partial removal of the clitoris. They also involved enlarging the vaginal opening. Many of the girls and women with CAH, in fact, underwent additional surgeries either to correct problems caused by earlier surgeries (including incontinence and fistulas), or because the earlier surgery failed to achieve "genital normalization." The "normalization" sought involved "appropriate appearance" and that penal-vaginal penetration was possible.

Ehrhardt's and Meyer-Bahlburg' s data not only included what they took to be the effects of androgens on the behaviors and temperament of girls and women with CAH, but also the results of experiments on laboratory rats that we earlier discussed that linked exposure to testosterone to behaviors and temperament associated with males. Ehrhardt and Meyer-Bahlburg reported that a much higher percentage of girls with CAH as compared with girls without the condition exhibited what they called "tomboyism." "Tomboyism" was operationally defined as exhibiting preferences for male rather than female playmates and active outdoor play, greater interest in careers rather than in becoming housewives or mothers, and less rehearsal of activities associated with mothering than is common among girls without the condition. The description of such girls as exhibiting "tomboyism" was based not only on observations made by Ehrhardt and Meyer-Bahlburg, but also on their interviews of parents, unaffected female siblings, and teachers, as well as on how the girls described themselves.

Feminist Critiques and Contributions

We have noted in the 1980s and early 1990s a central focus of feminist critiques of brain organizer hypotheses was their linear, hierarchical form (e.g., Longino 1990). In terms of the hypotheses about behavior and temperament now under consideration, feminists questioned the simplicity of the explanations given then current knowledge of the role of non-biological factors in affecting the behaviors and temperaments the hypotheses sought to explain; the appropriateness of extrapolations of research results involving much simpler species to human behavior and temperament; and they argued that background assumptions reflecting social beliefs about sex/gender, including gender stereotypes and evaluatively thick concepts, mediated the distance or gap between the hypotheses outlined above, and the data taken as evidence for them. We consider their critiques of specific hypotheses in the order in which they are presented above.

Feminist biologists argued that those offering brain organizer hypotheses to explain "male aggression" failed to operationalize either "aggression" or "aggressivity" in terms of a set of specific behaviors or temperamental traits that justified attributing either or both to males of many species, including human males. Rather, they argued, the behavioral criteria cited varied with the species of laboratory animal studied (Bleier 1984). In rats, for example, the actual phenomenon often referred to as "aggressive," and studied and measured for its frequency, was fighting behavior between two caged animals – not, as biologist Ruth Bleier pointed out, "attack behavior or the actual initiation of fighting by one of the two animals" (ibid., 95). But, she argued, "almost without exception," the titles of papers and their conclusions focusing on the phenomenon "used the term *aggression* or *aggressivity*, rather than language describing the phenomena actually observed, for example '*fighting encounters*' in caged animals" (ibid., 95; emphasis in original). In addition, feminists argued that because many of the animals used to support generalizations about aggression among males are caged and in artificial settings, it is far from clear that the behaviors so observed are characteristic of conspecifics in the wild. Feminists pointed to what they took to be another problem with such extrapolations. When aggression was attributed to men, it was not limited to actual fighting behavior; rather, its attribution was frequently based on men's perceived assertiveness and leadership qualities, characteristics not attributed to male laboratory animals.

Feminists also argued that explanations proposing unidirectional causal relationships between exposure to pre- and postnatal testosterone and aggression do not take into account that behavior and environmental factors affect hormone levels (e.g., Longino 1990). They noted that even in much simpler species, laboratory studies report relationships between environmental factors and hormone levels. Moreover, feminists argued, given appropriate stimuli, fighting behaviors can be elicited in any species or sex. As Bleier argued, "Female hamsters, gerbils, and vervet monkeys fight more than the males of their species (Floody and Pfaff 1974)" (Bleier 1984). In short, feminists argued, hormone levels and aggressive behaviors, even in much simpler species, are often both species-specific and context-specific. Gender stereotypes linking men with aggression, and evaluatively thick concepts such as "male aggression," feminists argued, led researchers to ignore the variability, species-specific, and context-dependent nature of behaviors lumped together under the category.

Finally, feminists pointed out that what researchers once took to be levels of aggression caused by androgens in laboratory rodents are now understood to be caused by estradiol (the strongest form of estrogen), to which androgens are converted when taken up by nerve cells.

Feminists also criticized Ehrhardt and Meyer-Bahlburg' s hypothesis that higher levels of exposure to androgens among girls and women with CAH caused sex-inappropriate behavior and temperament (e.g., Longino and Doell 1983; Longino 1990). They questioned the scientific status of the notion of "tomboyism" and argued that it, and attributions of it to those with CAH, reflected gender stereotypes. They also criticized the acceptance at face value of the observations made by parents, siblings, and teachers of such girls, given that they were aware of the girls' condition and their observations might have been shaped by preconceptions about how "normal" boys and girls behave. And, feminists suggested, an awareness that one's genitals are viewed as "abnormal" which is reinforced by medical interventions might also help explain the behavior and self-perceptions of girls with CAH. (e.g., Longino 1990).

In summary, feminist criticisms of explanations of sex/gender differences in human behavior that took them to result from pre- and postnatal exposure to testosterone cite the very considerable distance between these hypotheses and data as compared to the results of hormonal manipulation of laboratory rats. They argue that the distance is in fact filled or bridged by unrecognized and unwarranted background assumptions, such as gender stereotypes and evaluatively thick, but not operationalized, concepts such as "aggression." An empirically adequate explanation of observed sex/gender differences in behavior or temperament, feminists argue, would include many social factors of the sort noted above that linear, hierarchical explanations do not take into account (Longino 1990).

Sex/Gender Differences in Heterosexuality

Beginning in the 1960s, differences in pre- and postnatal exposure to androgens in human males and females were proposed to explain differences between "male heterosexuality" and "female heterosexuality." At the time, sexologists and other scientists took "male sexuality" and "female sexuality" to be mutually exclusive. The hypothesis we consider, offered by Ehrhardt and her colleagues, K. Evers and John Money, focused on the "sexuality" and "sexual behavior" of girls and women with CAH. They proposed that their higher than normal exposure to androgens during fetal development explained why they exhibited behaviors and preferences that, from the 1960s through the 1980s, were taken to be characteristic of "male heterosexuality" and uncharacteristic of "female heterosexuality" (Ehrhardt et al. 1968). The model of heterosexuality that Ehrhardt and her colleagues assumed proposed that the major differences between men's and women's heterosexuality involve differences in "libido." "Libido" was operationally defined in terms of several preferences: the number of sexual partners an individual was interested in having; a preference for 'versatility' or a preference for 'conservatism' in terms of sexual positions; being passive and receptive, versus being active and initiating; frequency of masturbation; and being "orgasmic" – taking orgasm to be the important aspect of heterosexual sex – versus "non-orgasmic" – valuing things like intimacy as much or more than orgasm (Jordan-Young 2010, 120).

The model accepted at the time that Ehrhardt and her colleagues undertook their studies proposed that men and boys have stronger libidos, as evidenced by what at the time was taken to be their greater interest in having multiple partners, in trying out a variety of sexual positions, and in achieving orgasm. They were also taken to engage more frequently in masturbation, and to be active in pursuing sex rather than just receptive to it. In addition, "male sexuality," unlike "female sexuality," was taken to include little by way of a "sentimental component" (ibid., 120). Together, these characteristics were taken as defining "male sexuality" – again, among heterosexuals – and the different, and opposing characteristics mentioned in the previous paragraph were taken to define "female sexuality." Until the 1990s, none of the characteristics associated with "male sexuality" in this early model were associated with "female sexuality" – but that has changed.

As noted, based on their studies of girls and women with CAH, which consisted for the most part of the descriptions the subjects provided and interviews of therapists treating them, Ehrhardt and her colleagues concluded that girls and women with CAH exhibit "male" rather than "female" sexuality, and attributed this to higher than normal levels of androgen exposure during fetal development. The evidence they cited included that the girls and women, and their therapists, reported that they frequently engaged in masturbation, and that their sexuality was not characterized by a sentimental component (Ehrhardt et al. 1968, 120).

Feminist Critiques and Contributions

Feminists have offered several responses to this hypothesis. One involves proposing alternative explanations of the sexual preferences girls and women with CAH and their therapists reported. As we earlier noted, girls with CAH typically undergo surgery to enlarge their vaginal opening. The girls were encouraged by medical personnel and therapists to engage in frequent masturbation (typically involving inserting one's fingers into the vaginal opening) to maintain the vaginal opening and flexible genitalia – again, to enable penile-vaginal penetration. From this perspective, the girls' and women's frequent masturbation is a function not of their sexuality, but of the fact that they were encouraged to masturbate in order to achieve or maintain what was taken at the time to be "normalization" of their genitalia. In addition, given partial or complete removal of the clitoris, which was common when Ehrhardt and colleagues were studying them, girls and women with CAH reported little or no clitoral sensation – so that achieving orgasm by masturbating (which, at the time, was associated with "male sexuality") was, in some cases, unlikely and impossible in others (Crouch et al. 2004). Nor, feminists argued, were reports that girls and women with CAH were "less sentimental" about sex surprising (recall, sentimentality was associated with "female sexuality"). Feminists cited their self-perceptions of their genitalia, and the reactions of others to them, as just as likely an explanation of their attitudes toward sex. So, feminists argued, in terms of some behavioral and temperamental characteristics associated at the time with "male sexuality" taken to be exhibited by girls and women with CAH, they can be equally well explained by factors other than higher than normal levels of prenatal androgen.

Feminists also point out that subsequent studies challenge the findings that led Ehrhardt and her colleagues to conclude that girls and women with CAH exhibited "male" rather than "female" heterosexuality as each was then defined (Jordan-Young 2010). One long-term study reported that many girls and women with CAH reported experiencing pain during intercourse and more than 30 percent reported never having engaged in it. As noted earlier, being interested in initiating sex is a characteristic associated with "male sexuality." But many girls and women with CAH said that they were not interested in initiating sex (Gastaud et al. 2007).

The differences between the findings of subsequent studies and those reported by Ehrhardt et al. have led some feminists to ask whether Ehrhardt and her colleagues' *expectations* about the effects of prenatal androgens on sexuality contributed to the conclusions she and her colleague reached about girls and women with CAH (e.g., Jordan-Young 2010). In Chapter 2 we discussed arguments offered by Hanson and others that observations are "theory laden," and that expectations are among the factors that contribute to what scientists, and all of us, observe.

More Philosophical Issues

"Quasi Experiments" in BOR

As noted when we introduced the issue of "quasi" versus "true" experiments, there are ethical and practical reasons why studies of human subjects can at best constitute "quasi-experiments," rather than "true experiments," as scientists understand the latter. The studies of human subjects we have considered do not engage in social engineering to control for all the variables that might contribute to mathematical abilities, behavior, temperament, or sexuality. Nor do they manipulate the hormone levels of these subjects, as experiments on laboratory rodents often do.

Feminist neuroscientist Rebecca Jordan-Young describes the difference between "true experiments" and "quasi experiments" as it relates to brain organizer hypotheses. "In true experiments of a brain organizer hypothesis," she points out,

> subjects would be randomly assigned to receive particular hormonal exposures, and their development would be observed over the life span, keeping rearing experiences and environments constant across experimental and control groups. Of course, this isn't possible with humans, so scientists must piece together evidence from animal studies, and from individual human quasi experiments that are by definition partial and uncontrolled. (ibid., 3)

In other words, Jordan-Young argues, "each study of how prenatal hormones sexually 'organize' the [human] brain" is not a true experiment" (ibid., 3). In addition, she argues, that consistency with other research findings shapes "the interpretation of every quasi experiment" as does seeing how well they all fit together (ibid.). They are not and could not be based on "true experiments" involving human subjects.

Again, it is not the reliance on or appeal to other research and hypotheses that is problematic, for these are common and often successful. Indeed, our discussions of feminists' critiques of various hypotheses and research programs indicate that feminists often themselves make such appeals. What feminists argue is at issue in this case is this. Many of the brain organizer hypotheses they criticize appeal to studies of laboratory rodents; some to hypotheses in empirical psychology about sex/gender differences in lateralization; and some appeal to data they take to be evidence of sex/gender differences in specific traits. They rely on what feminists argue are *gender stereotypes* and *evaluatively thick concepts,* such as those associated with "tomboyism."

Is Evolution the Ultimate Explanation of Sex/Gender Brain Differences?

Jordan Young undertook an extensive analysis of over 300 research studies in BOR, and found that many of the researchers involved in them take there to be fundamental parallels between differences in sex organs related to sexual reproduction, on the one hand, and differences between the brains of males and females, on the other hand. Using a vivid analogy, Jordan-Young argues that BOR assumes "that the brain is a sort of accessory reproductive organ." She takes this hypothesis to be what she refers to as this "very simple idea," namely that

> Males and females don't just need different genitals in order to have sex, or different gonads that make the eggs and sperm necessary for conception. Males and females also need different brains so they are predisposed to complementary sexual desires and behaviors that lead to reproduction. [Brain Organization Research] suggests that the same mechanism is responsible for both kinds of development – that is, for sexual differentiation of "both sets" of reproductive organs: the genitals and the brain. (ibid., 21)

Many sex/gender differences proposed in BOR are virtually the same as those that many evolutionary theorists take to be the result of natural or sexual selection because they were conducive to reproductive success during the Pleistocene. So, it is likely that at least some evolutionary theorists assume that evolutionary theory provides "the ultimate explanation" of such differences: that natural or sexual selection "selected for" the genes causally implicated in the differential exposure of male and female fetuses to sex hormones. After all, such differential exposure does result in the anatomical differences that enable sexual reproduction. And if Jordan-Young is correct, BOR assumes that brains are auxiliary reproductive organs, equally important to reproductive success.

So, we need to ask if one or another version of evolutionary theory does provide the ultimate explanation of what brain organizer hypotheses propose are sex/gender differences in cognition, behavior, and temperament related to sex/gender differences in brains? Given that reproductive success is necessary for evolution to occur, feminists readily grant that attention to factors hypothesized as contributing to it is clearly warranted. But feminists' arguments suggest that, as is usually the case, the devil is in the details.

As we noted in our discussion of Human Sociobiology in Chapter 3, many evolutionary theorists assume that "a sexual division of labor" was a key characteristic of life in the Pleistocene during which, it is assumed, many traits unique to humans emerged. They hypothesize that our male ancestors were responsible for providing for and protecting our female ancestors and children, and that our female ancestors were primarily involved in activities associated with reproduction and child rearing (e.g., Wilson 1978). Those who propose this division in labor take it to be the result of selection pressures and to reflect the selection of "complementary" sex/gender differences conducive to survival and reproductive success. But, feminists argue, there are problems with this line of reasoning.

One is that there isn't evidence that life in the Pleistocene was organized in the ways just described. Rather, feminists argue as we have noted that, the reconstructions of life during the period being appealed to reflect an unwarranted imposition of assumed features of contemporary cultures to past cultures. In addition, we considered what feminists argue are androcentric accounts of contemporary hunter-gatherer groups and of non-human primate behavior that such reconstructions rely on. An additional problem, feminists argue, is that in many respects the sexual division of labor that does exist in contemporary cultures is changing in many societies. Evidence has also emerged that indicates that gender roles and divisions of labor by sex/gender long assumed to characterize life in ancient groups and civilizations, were far from stable. Rather gender roles and divisions of labor by sex/gender varied substantially, not just by group, but also over time. This evidence emerged as a result of relatively recent and robust interest in archaeology, among feminists and non-feminists, in gender and gender relations in ancient groups and civilizations. Researchers have found that changes in both apparently occurred, for example, when a group or civilization was conquered by another, and previously non-gender dimorphic divisions in labor gave way to gender-based divisions that were common in the now dominant group. In short, this research strongly suggests that gender roles and sexual divisions of labor are neither stable nor universal despite longstanding assumptions by researchers, including Human Sociobiologists, to the contrary (Wylie 1997).

Additional issues raise questions about evolutionary explanations of sex/ gender differences in cognitive abilities, behavior, temperament, and (heterosexual) sexuality. Darwin, Human Sociobiologists, and some other evolutionary theorists typically assume that what have been taken to be significant parallels in sex and sex/gender differences between many species constitutes *prima facie* evidence that the sex/gender differences assumed or argued for are evolutionary in origin. But times are changing. We have seen, for example, how the so-called primate pattern in terms of male and female behavior widely accepted in the 1970s had to be abandoned when detailed observations of the females of several primate species challenged its core hypotheses that males are dominant and aggressive, that females are submissive and non-aggressive, and that males, rather than females, initiate sexual encounters. In light of these findings, many traditional extrapolations to humans of what were assumed to be sex differences in species closely related to us, do not hold up. So, what are the implications of these findings for evolutionary hypotheses about sex/gender differences?

Obviously, these are complex and thought-provoking issues. We return to them after we discuss the explanations Evolutionary Psychologists propose of sex/gender differences in mating and parenting strategies in the next chapter.

Conclusion

We have not asked, as we did of primatology and developmental biology and other fields, whether feminism has made a difference to research that has focused on identifying and explaining differences between "male brains" and "female brains." It is time to do so.

Jordan-Young's analysis of over 300 studies in BOR seeking to identify (or claiming to identify) differences in "male" and "female" brains, including those of men and women, suggest that, as of now, it has not. And a recent anthology, *Neurofeminism*, whose editors and contributors include cognitive psychologists, neuroscientists, behavioral biologists, and philosophers, also does not suggest that the extensive critiques feminists within the neurosciences have offered of brain organizer research, have yet resulted in changes in the theoretical or methodological assumptions underlying such research (Bluhm et al. 2012).

This may be a function of the deep-seated and long-standing assumption that men's and women's brains are different. That such differences are real and obvious has been maintained by philosophers, theologians, and many scientists. This view dates to the Ancient Greeks, was advocated by Darwin, and accepted by many anthropologists and psychologists in the nineteenth and twentieth centuries. Alternatively, or in addition, the lack of change may be a function of the relatively more recent entrance into neuroscience and related disciplines of scientists who have brought feminist perspectives to bear on their fields, compared with what occurred in primatology and developmental biology.

Yet, there are reasons to think that feminists' engagements with research priorities and questions about sex/gender in areas of neuroscience may well come to have an impact. For one thing, the number of articles, books, and anthologies published by feminist neuroscientists and science scholars arguing for changes in the field, while more recent than those considered in previous chapters, is now substantial and growing. For another, many of the arguments feminists are advancing strongly parallel those feminist scientists offered in fields that did undergo changes in terms of understandings of and approaches to gender. Perhaps as importantly, many feminist arguments for change are as positive in their recommendations as they are critical, and in this way also parallel the changes feminists in developmental biology and primatology recommended and that eventually took hold. Feminist neuroscientists and science scholars have also organized international conferences devoted to gender and neuroscience that were well attended and resulted in publications of books and special issues of journals. And there is a growing number of international networks devoted to bringing changes to research priorities and methodologies in neuroscience related to sex/gender.

Finally, although neither the members of neurofeminism networks, nor those who have attended the conferences mentioned above, are monolithic in terms of their views, there are two recurring and important themes in their presentations and writings. One is the highlighting of the ethical implications of research in the neurosciences, including that devoted to sex/gender. Members of these networks call on neuroscientists to become more aware of the ethical implications of their hypotheses, including those they advocate in publications written for the lay public. Other themes involve epistemological issues. There are feminist arguments, for example, that neuroscientists must come to acknowledge their situatedness in terms of specific scientific and social contexts, and its potential effects on their hypotheses and research priorities. Additional arguments that parallel those offered by scientists representing a variety of fields and emphasizing a variety of topics, emphasize the need for "socially-responsible science." (Many contributors to the collection *Neurofeminism*, edited by Bluhm, et al. (2012) discuss one or more of these issues). We briefly discuss the extensive and growing interest in "socially-responsible science" in Chapter 9.

So, although the kinds of change that have come about in some sciences in terms of approaches to sex/gender have not yet come about in the neurosciences (with the exception of feminist work therein), there are reasons to think that feminist engagements in and with these sciences will eventually have an impact.

8   Evolutionary Psychology

Introduction

Evolutionary Psychology posits features of human psychology that it maintains are the products of evolution, having been selected for in the Pleistocene and inherited by modern humans. Most of its practitioners are trained in psychology and/or cognitive science, rather than biology. We consider it in this text because evolutionary psychologists maintain that their hypotheses about human psychology, including about sex/gender differences, are based, in specific and substantive ways, on hypotheses and methods of evolutionary biology. To make Evolutionary Psychology's relationships to evolutionary theorizing clear, we begin with some background, including reminders of discussions in Chapters 2 and 3 in which we considered evolutionary theorizing.

We have noted that most evolutionary biologists view natural selection as an important – albeit, not the only – mechanism or factor that determines the directions and results of evolution. Researchers in Evolutionary Psychology (hereafter EP) consistently appeal to natural selection in positing, explaining, and predicting features of human psychology that, they maintain, influence human behavior. We have also considered Darwin's arguments for sexual selection and the development in the twentieth century of Parental Investment Theory (PIT). Evolutionary psychologists' descriptions, explanations, and predictions of human mating and parenting strategies, particularly differences between women's and men's strategies, consistently appeal to sexual selection and PIT. Despite the critiques of each theory we considered in Chapters 2, 3, and 4, evolutionary psychologists describe both as central to mainstream evolutionary biology.

In Chapter 3, we also briefly discussed Human Sociobiology (HS) as it was initially proposed by geneticist E.O. Wilson and developed by theorists in the 1970s and 1980s. The stated goal of HS was identifying genes selected for during the Pleistocene era that continue to determine human social behavior and forms of social organization. We noted that this project came to be viewed by many biologists and philosophers of biology as unsuccessful, given that few, if any, of the genes HS posited were identified.

However, although HS was not without its critics – and, as we noted, some of the most severe were other geneticists and evolutionary theorists – many scientists and science scholars continue to view the *general* project that Wilson advocated – seeking to explain human behavior and forms of social organization on the basis of evolution – to be in keeping with the goal of an integrated and, to the extent possible, unified science. That is, for many the intuition remains, and it is a strong intuition, that given that humans evolved, it is eminently reasonable to assume that our behavior and psychology reflect to some extent the effects of natural selection, as well as other evolutionary mechanisms and forces.

It is at the intersections of the history of projects such as Darwin's and HS, and the intuitions just mentioned, that some psychologists began to develop EP in the late 1980s and 1990s. Architects of the discipline emphasize "conceptual integration" as providing the rationale for the research program. Eschewing a direct causal link between one or more genes and a specific social behavior, EP posits causal relationships between, on the one hand, selection pressures they take our ancestors to have faced during the Pleistocene, and, on the other hand, innate cognitive mechanisms and "predispositions" that we have inherited. Its practitioners do not eschew evolutionary explanations of human social behavior and forms of social organization. Rather, they maintain that the cognitive mechanisms and predispositions they identify are what contribute to these phenomena. In other words, EP posits an intermediary between genes and behavior – namely features of human psychology selected for in the Pleistocene.

The introduction of cognitive mechanisms and psychological dispositions as having an intermediary causal role in terms of human behavior does not mean that EP differs in all respects from HS. In their introduction to the anthology *The Adapted Mind: Evolutionary Psychology and the Generation of*

Culture, the editors of the volume who are also architects of the discipline, describe the core assumptions of the research program as follows:

> The central premise of *The Adapted Mind* is that there is a universal human nature ... that exists primarily at the level of evolved psychological mechanisms, not of expressed cultural behaviors ... A second premise is that these evolved psychological mechanisms are adaptations, constructed by natural selection over evolutionary time ... A third assumption ... is that the evolved structure of the human mind is adapted to the way of life of Pleistocene hunter/gatherers. (Cosmides et al. 1992, 5)

This passage clearly indicates that several core assumptions of HS are carried over to EP. These include that there is a basic, universal human nature; that problems facing our ancestors in the Pleistocene provided the selection pressures that created that nature; and that the cognitive traits and psychological dispositions that EP proposes to have resulted, are "adaptations." As we later discuss, HS and EP also share a methodological commitment to a view that has come to be called "adaptationism": that traits should be assumed to be the product of natural and/or sexual selection unless shown otherwise.

Other significant parallels between the research programs include that EP, like HS, uses "reverse engineering," a method we later discuss, to 1) identify historical selection pressures on the basis of which they *predict* aspects of human nature, and 2) *to explain* apparently confirmed aspects of human psychology using reconstructions of ancestral life in the Pleistocene. Like HS, EP also relies on sexual selection and PIT. And the sex/gender differences proposed or assumed in the two programs are often the same. What differs is how human behavior is explained. HS posits specific genes as the cause of specific behaviors. EP posits cognitive mechanisms and psychological predispositions produced by evolution that, given appropriate triggering conditions, result in specific human behaviors. We will see that the shift to cognitive mechanisms and predispositions that can but may not be "triggered," allows EP to accommodate cross-cultural differences that HS could not.

So, in terms of EP's appeals to what it maintains are features of "mainstream" evolutionary biology (a claim we will have reason to evaluate), and the strong parallels between its claims about sex/gender and those of Darwin, PIT, and HS, it seems appropriate to treat EP as a *hybrid* discipline. Its core features include empirical and methodological commitments characterizing some (but not all) approaches in evolutionary biology, as well as some that (in their general outlines, if not in their details) are common in cognitive science.

Basic Commitments of Evolutionary Psychology

As we have noted, the authors of the introduction to *The Adapted Mind* (hereafter, TAM) describe Evolutionary Psychology as a research program that seeks to identify, and explain or predict, evolved psychological mechanisms selected for in the Pleistocene and inherited by subsequent generations of humans. Frequently drawing an analogy between the human mind and a Swiss army knife – a small knife that includes several blades as well as other small tools – its practitioners argue against the hypothesis that the mind is endowed with something like an all-purpose problem-solving mechanism. Rather, they contend, the human mind is endowed with numerous, problem- or domain-specific, "cognitive modules" (sometimes called "cognitive mechanisms," "predispositions," and/or "Darwinian modules"). In general, "the modularity" of the human mind is a common hypothesis in Cognitive Science.

As EP theorists speak of them, cognitive modules are something like software programs or algorithms selected for to solve specific adaptive problems they take our ancestors to have faced. As noted above, some evolutionary psychologists speak of cognitive modules and/or mechanisms, others of predispositions. We don't need to try to get to the bottom of why there are different locutions, because the specified modules, mechanisms, and predispositions function in the same way in the explanations, predictions, and arguments that evolutionary psychologists advance. They are inherited, innate, and universal features of human psychology – and they are adaptations.

As we noted in earlier chapters, in widely accepted usage in evolutionary theorizing (if not always adhered to in practice as critics often point out), an "adaptation" is a trait that has been "selected for" by natural or sexual selection. It is not just "adaptive" – i.e., it is not just conducive to an organism's survival or successful reproduction (e.g., Sober 1993). Evolutionary psychologists accept this distinction and often state that a trait selected for during the Pleistocene may come to be *maladaptive* in a later environment. For example, evolutionary psychologists who advance the hypothesis that a predisposition to commit rape is an adaptation linked to reproductive success, or a byproduct of an adaptation leading men to seek multiple mates, also state that, whatever caused the selection of the predisposition in ancestral contexts, the predisposition is now decidedly maladaptive. Parallel arguments have been offered about other cognitive traits taken to be selected for during the Pleistocene, including a predisposition in men to aggressiveness and sexual jealousy. We later explore whether descriptions of such traits as "maladaptive," use that term in ways in keeping with how it is understood in evolutionary biology.

As I have argued in another place, what theorists in EP often do not take into account, although Darwin did and evolutionary biologists continue to, is the phenomenon Darwin called "conversion of original function." This involves cases in which a trait's current function is different from that for which it was originally selected. A commonly cited example is the original function of feathers among ancestors of birds that did not fly. EP appears to be committed to a straightforward (non-branching, non-changing) transmission to contemporary humans of psychological features selected for in the Pleistocene (Nelson 2003). If this is correct, arguably its approach to adaptation differs from how Darwin and many contemporary evolutionary biologists approach it.

We can now identify some important differences between EP and HS. As briefly mentioned above, because EP emphasizes "predispositions" and other features of human psychology, it is compatible with the significant historical and cross-cultural differences in social beliefs, practices, and forms of social organization, with which HS's purportedly genetically determined "cultural universals" were not. Hypotheses positing psychological *predispositions* and modules allow EP to incorporate a notion of "triggering circumstances" in the absence of which some behavior linked to an innate predisposition may not be exhibited.

One might view the hypothesis of triggering circumstances as allowing for hypotheses that are, in principle, unfalsifiable (and, thus, some would argue, unscientific). But an emphasis on contingency is well in keeping with evolutionary theorizing. In addition to triggering circumstances, researchers in EP also maintain the more general hypothesis that neither genes nor psychological mechanisms alone determine human behavior. Many other kinds of factor, they maintain, can contribute to whether an evolved psycho-logical mechanism or predisposition is expressed behaviorally (e.g., Buss 1999), although they often do not discuss them in detail. So, too, the large number of psychological mechanisms and predispositions the program posits are argued to allow for a good deal of flexibility on the part of humans (e.g., Buss 1999).

These features of EP suggest that the important empirical questions to be asked concern the nature and strength of the evidence for the existence of the evolved psychological mechanisms and predispositions its practitioners posit, *and* the nature and strength of the evidence that they are in fact *adaptations* (i.e., the products of natural or sexual selection).

Another difference between EP and HS is that those working in EP do not have to contend with the issue of how likely it is that there is, say, a "gene for promiscuity" or one for "coyness" as HS did. Given that the human genome only includes approximately 22,000 genes (less than an ear of corn), such gene specificity seems highly unlikely. Researchers in EP, however, do have to contend with criticisms leveled by cognitive scientists who, although they accept the hypothesis that there are cognitive modules, question whether the human mind is or could be endowed with as many psychological modules or mechanisms as EP maintains it is, and question the evidence that supports such claims (Bechtel 2003).

Methods Used in Evolutionary Psychology

Evolutionary theory, like physical anthropology, geology, and archaeology, is a historical science, which means that theorists reason from what can be observed now (in the case of human evolution these include artifacts, fossils, and the like) to past phenomena that cannot now be observed. It is generally recognized by evolutionary biologists and philosophers of biology that there are two general ways to arrive at an evolutionary (i.e., adaptation) explanation.

One of these ways is to make use of *history*: concrete knowledge about relevant historical facts – including information about variations in an ancestral species in which, it is thought, the adaptation emerged; and about specific selection pressures with which that species was faced that explain the selection of the variation in question. There are, in principle, two ways evolutionary theorists might obtain the needed historical facts. The most direct method of this kind makes use of population genetics, which studies genetic differences within and between populations, including ancestral populations. But, as philosopher Robert Richardson and others make clear, it is not currently feasible to use population genetics to explain the traits of many species, and certainly not possible in the case of many human traits, as adaptations (Richardson 2001). A less direct historical method, the comparative method, uses knowledge about phenotypic traits – the observable char acteristics of organisms, rather than genotypes, to identify variations in an ancestral species (using, for example, the fossil record) that, together with information about then current selection pressures, can be used to make a plausible case that a trait is an adaptation. But Richardson, among others, notes that because there are significant gaps in our knowledge of human phylogeny, as well as gaps in other knowledge the method makes us of, we cannot now use the comparative method to reconstruct the selection of human traits. In Richardson's words:

> The [comparative] method imposes severe requirements: we need comparative data on related taxa, developmental information, information concerning the character of the environment, the trait family under consideration, and the relative adaptiveness of the traits characteristic of the several taxa. (ibid., 336)

And, as Richardson notes, there are formidable problems confronting any effort to exploit resources of the biological sciences (e.g., in phylogenetic systematics) to get such information. They include that the relationships between three of four hominid species remain unclear and that more than one phylogeny enjoys a degree of acceptance.

The second general way to arrive at adaptation explanations is by using "reverse engineering." Reverse engineering is, by its nature, an *ahistorical* method of explanation. Beginning from the assumption that a given phenomenon is the result of some process of building or design, one infers *why* it is the way it is and/or *why* it came to be (Dennett 1995). In evolutionary theorizing reverse engineering is ahistorical because it does not make use either of knowledge about phenotypic variations in the ancestral population from which a trait was "selected for" or of information about the actual selection pressures facing that population. Nor does it make use of population genetics. Reverse engineering in evolutionary theorizing is uncontroversial when history is irrelevant – for example, when it is used to derive an explanation of a trait's present adaptiveness. But when it is used to generate historical explanations – for example, when ancestral conditions and causal processes are inferred from present traits or capacities – it is controversial. As reverse engineering in evolutionary theorizing enjoys advocates as well as critics, we later consider influential arguments for each view.

Here we focus on the two methods that the editors of TAM identify as the "most direct methods" researchers in EP are able to use to establish the "adaptive function" of an evolved psychological mechanism. Both, we will see, make use of reverse engineering. One line of reasoning involves identifying historical selection pressures facing our ancestors in the Pleistocene using what evolutionary psychologists claim is known about current human psychology. Cosmides et al. describe this line of reasoning:

> Researchers can start with a known psychological phenomenon, and begin to investigate its adaptive function, if any, by placing it in the context of hunter/ gatherer life and known selection pressures.. . . . to try to understand what its adaptive function was – why that design was selected for rather than alternative ones. (Cosmides et al 1992, 10)

A hypothesis proposed by Leda Cosmides and John Tooby, that human minds are endowed with "a cheater-detection" mechanism, involves such reasoning (Cosmides and Tooby 1992). Their reasoning begins from experiments undertaken by others in which a majority of subjects given what is basically the same logic puzzle (in terms of its set up, possible solutions, and correct solution), consistently do less well in solving the puzzle when it is presented in abstract logical terms than they do when it is presented as a puzzle involving the detection of cheaters.

Figures 8.1 and 8.2 illustrate a version of the "two" puzzles. Figure 8.1 is presented first to subjects, who are told that each of the 4 cards has a letter on one side and a numeral on the other side. They are then asked "Which of these 4 cards do we need to turn over to be sure that the generalization 'All cards with a D on one side have a 3 on the other side' holds in terms of these

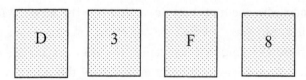

Figure 8.1 Cheater detection illustration 1.

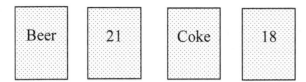

Figure 8.2 Cheater detection illustration 2.

cards?" Many choose too many cards (including, for example, the card marked 3, even though the generalization does not entail that a card with 3 on one side has a D on the other side).

They are then presented with the set of cards illustrated in Figure 8.2.

Subjects are told that each card has the age of a person sitting at a bar on one side and what they are drinking on the other side. They are asked "Which of these 4 cards do we need to turn over to be sure that the rule 'No one under 21 may drink beer here' is being followed?"

In both "puzzles" or cases it is the first and the fourth cards that need to be turned over Experiments show that many who fail to solve the first "puzzle," succeed in solving the second. This difference is taken by Tooby and Cosmides to be the kind of phenomenon that requires an evolutionary explanation. We turn now to a brief explanation of why.

Although doubts have been raised about the accuracy of the oft-told explanation of the QWERTY keyboard – which maintains that it was designed to prevent fast typists using old-fashioned typewriters from locking keys of frequently used letters – it is often used as an example of the kind of object (read "trait" as we're discussing evolution) that is the way it is due to contingent historical factors. An argument offered by Stephen Jay Gould illustrates one reason some evolutionary theorists take traits that Daniel Dennett describes as "QWERTY-like" phenomena to lend themselves to adaptation explanation (Dennett 1995). Gould used the panda's "thumb" to argue that traits that come about through the "jury-rigging of available parts" are the best evidence of evolution. It turns out that the panda's so-called thumb, which enables it to hold bamboo in one paw and strip parts to eat using the other, isn't a digit at all. It is an enlarged wrist bone that, together with the development of related muscles, resulted in what appears to be a sixth digit. Comparing anatomical features of pandas and ordinary bears, Gould argued that the panda's "thumb" is an example of the kind of jury-rigging of available parts that serve as evidence of evolution and challenge the view that organisms are the product of perfect "design" (Gould 1980).

Cosmides and Tooby take the inconsistency in people's reasoning revealed in the experiment just outlined to indicate historical contingency. To explain the ability to solve a logic puzzle presented in terms of detecting cheaters as an adaptation, they use reconstructions of the Pleistocene we have discussed to identify the selection pressures that led to the "selection for a 'cheater-detector'" mechanism. They assume that what hunting and gathering provided was shared among the group. Based on this assumption, they reason that the ability to detect cheaters – those who did not contribute in some way to the processes, but did attempt to partake of the results, would be important to maintaining a system of social exchange.

Of course, one can note that there are different ways that people cheat, and many ways they might be caught doing so. But Cosmides' and Tooby's can argue that despite how the adaptation originally arose, a cheater-detection module can be and is used in a variety of other contexts. We later consider an explanation for the inconsistencies in solving the "two" puzzles that is an alternative to their explanation.

We also later consider hypotheses about sex/gender offered in EP that make use of reverse engineering to identify psychological mechanisms and predispositions selected for in the Pleistocene and passed down to us. Here we note three things feminists criticize about hypotheses about sex/gender differences arrived at using this method: 1) as noted in several earlier chapters, the reconstruction of Pleistocene life is itself arrived at using reverse engineering that is based on mid-twentieth century accounts of contemporary hunter-gatherer groups; 2) those ethnographic accounts and the reconstructions of ancestral life based on them are now taken to be seriously flawed by many anthropologists; and 3) feminists' and others' critiques are now widely taken to have shown that both early ethnographic accounts of contemporary hunter-gather groups and the reconstructions of ancestral life based on them, were androcentric and ethnocentric.

A second method the editors of TAM cite moves, so to speak, in the opposite direction: from the past to the present, and it leads to *predictions*, rather than explanations, of adaptations – specific psychological mechanisms or predispositions among contemporary humans resulting from selection pressures our ancestors faced in the Pleistocene. The editors of TAM describe the method, including how knowledge of the relevant historical selection pressures can be attained, this way:

> By combining *data from paleontology and hunter/gatherer studies* with principles drawn from evolutionary biology, one can develop a task analysis that defines the nature of the adaptive information-processing problem to be solved. (ibid., 11, emphasis added)

The solutions that result are, again, evolved psychological mechanisms. Because this method generates testable hypotheses predicting specific "information-processing mechanisms," Cosmides et al. maintain that "it is immune to the usual (but often vacuous) accusation of *post hoc* storytelling" (ibid., 11). That is to say, they maintain that because it is possible to *test* a prediction proposed in EP, that we can be assured that the hypothesis is not a "just so story." We will later see that critics of the use of reverse engineering in evolutionary theorizing, charge that the adaptation explanations that result often lack evidential support. They use the phrase "just so stories" to describe the hypotheses in question (e.g., Gould and Lewontin 1979). It is such charges to which the editors are responding.

As an example of this second method, consider the hypothesis advanced by David Buss (among others) that women have "an evolved psychological mechanism to prefer men with resources" (in Buss 1999, among other places). It is instructive in several respects. Feminists point out that the "adaptive problems" surrounding mating and parenting constitute only a subset of those impacting survival and reproductive success. But evolutionary psychologists, like human sociobiologists, devote a lot of attention to identifying sex/gender differences in such strategies.

Buss maintains that his hypothesis is derived from knowledge of an adaptive problem Pleistocene women faced: the need to find a mate who would provide for them and their offspring. This "adaptive problem" in turn is identified, Buss maintains, based on what we "know" about contemporary hunter-gatherer groups. But, as we have noted, both the accounts of contemporary hunter-gather groups, and the reconstruction of ancestral life based on them, are controversial. So, too, the hypotheses from paleontology and archaeology on which EP relies are now taken by many archaeologists to be dated. In Chapter 7, we discussed some of the factors that led many archaeologists to conclude that the sex/gender relations in ancient populations that EP assumes were neither universal nor stable.

But recall that Cosmides et al. maintain that the method Buss is using yields testable predictions and, thus, such hypotheses are *empirically constrained*. After all, predictions can and often do fail. Buss and his colleagues did test the hypothesis by surveying women in a variety of cultures who differed from one another in many respects, including socioeconomic status. They presented women with a list of traits of potential mates and asked them to prioritize the traits. And, indeed, Buss reported, the survey did confirm the hypothesis – most of those surveyed did place economic resources high on the list of traits they valued in a potential mate. Buss claims that the confirmation of the preference among contemporary women demonstrates that these women "are the descendants of ... successful ancestors, and so have inherited their mate preferences" (Buss 1999, 111).

When we turn to critiques of EP offered by feminist and other scientists and science scholars, we will return to a point discussed in earlier chapters: namely, that the evidential warrant for a given hypothesis is not solely a function of its apparent confirmation, but also a function of the lack of alternative, equally viable explanations of the phenomenon. To anticipate the discussion of the relevance of this criterion to hypotheses advanced in EP, consider an alternative explanation for the preference Buss's surveys confirmed (one that many of my students propose): given that, in general, women tend to have fewer resources then men, it is entirely rational for women to prefer mates with more resources over those with fewer resources. As philosopher Daniel Dennett argues, and we later consider his argument for this point, there is no need of an adaptation explanation for behaviors and preferences that are obviously rational (Dennett 1995).

A similar point about alternative explanations can be made about Cosmides' and Tooby's hypothesis that we are endowed with a cheater-detection mechanism. An alternative explanation for why subjects did not correctly solve the puzzle presented in abstract, logical form (and, again, this is one my students often offer) is that many people are sure that they are incapable of solving "logic puzzles" so presented. Refiguring the puzzle into one involving familiar experiences may well make them less uneasy, and concrete rather than abstract puzzles may be, for many, quite a bit easier to solve. To test my students' hypothesis, the next time I presented the puzzles, I began with the second puzzle (involving drinking beer, being under 21, etc.). Their overall success in solving this form of the puzzle was in keeping with that of previous classes. But first providing the puzzle in its concrete form resulted in many more solving the puzzle in its abstract form than had done so in previous classes. This suggests the possibility that my students were onto something. The more important point is that there is an alternative explanation to that advanced by Cosmides and Tooby.

Sexual Selection, Parental Investment Theory, and Sex/Gender Differences in Evolutionary Psychology

As we have noted in several contexts, the "adaptive problems" surrounding mating and parenting constitute only a subset of factors that can have an impact on reproductive success. As we have also noted, practitioners of EP, like Human Sociobiologists, devote a lot of attention to these issues. A partial list of the cognitive mechanisms and/or predispositions proposed in EP includes:

- Women's preference for men with economic resources (Buss 1999).
- Men's predisposition to sexual jealousy – which, among other phenomena, is said to explain domestic violence and violence against stepchildren (Daly and Wilson 2005).
- Men's predisposition to find traits linked with fertility attractive (Singh 1993).
- Women's predisposition to impose more stringent standards than do men before they will consent to (heterosexual) sex (Buss and Schmitt 1993).
- Women's predisposition to prefer men willing to "invest" in them and their children (Buss 1999).
- A predisposition to rape in men that is either an evolutionary adaptation related to reproductive success, or a by-product of an adaptation that leads men to desire multiple mates (Thornhill and Palmer, 2000).

Each of these alleged predispositions was arrived at by using one of the two methods cited by the editors of TAM that we earlier noted. There are several features common to them. Whether beginning from a reconstruction of an adaptive problem hypothesized to have faced our ancestors in the Pleistocene, or from a generalization about an aspect of current human psychology, the proposed features of human psychology just noted, and the explanations EP offers of them, assume they are adaptations and they are arrived at through reverse engineering. Recall, however, that these reconstructions of life in the Pleistocene have been subjected to serious critique And the hypotheses all appeal to sexual selection and Parental Investment Theory, which we noted in Chapters 2, 3, and 4 are also controversial. Here, we discuss these issues in relation to two hypotheses listed above Later we return to the others.

We have already discussed how Buss's hypothesis that women have an evolved predisposition to prefer men with resources is of this kind. This is also the case in terms of hypotheses explaining domestic violence that are based on the hypothesis that men have a predisposition to sexual jealousy. Those positing this predisposition cite problems facing our male ancestors in insuring that they didn't unknowingly "invest" in offspring that weren't their own. They often begin from information provided by studies of spousal homicides and battery in which sexual jealousy is consistently found to be a factor. The adaptive problem evolutionary psychologists assume – the uncertainty of paternity among male ancestors – is derived from Parental Investment Theory, but also from what is uncontroversial – the uncertainty, until relatively recently, surrounding the question of paternity. That paternity has been a matter of concern for men is not in question. After all, Homer addresses it in *The Odyssey* when a disguised Athena asks Odysseus' now grown son, Telemachus, who his father is; the young man replies "Never yet did any man know, on his own, who gave him life." Only mothers (unless they were completely isolated from any other potential mates) knew for much of human history. The question to be asked, however, is whether there is empirical evidence that the "problem of paternity" resulted in an *adaptation* – an evolved predisposition to sexual jealousy in men.

Feminists have been highly critical of EP's hypotheses about sex differences in cognitive mechanisms and predispositions related to mating and parenting strategies. Before turning to a discussion of such critiques, we consider debates about two methodological assumptions at work in the program: adaptationism and the appropriateness of using reverse engineering to arrive at adaptation explanations.

Philosophical Issues

Adaptationism

As we have noted, explanations in evolutionary theorizing take a variety of forms because there are a number of processes and forces that have effects on the directions and results of evolution. In addition to natural selection and sexual selection (not all evolutionary theorists do assume sexual selection as we have noted), forces such as genetic drift, mutation, and migration, as well as developmental constraints, contribute to genotypes and phenotypes.

There are disagreements about the relative impact of these various mechanisms and forces, including disagreements about how much emphasis should be placed on natural selection and on adaptations. But, as many have pointed out, the most defensible view is that not all traits have been selected for – that is, not all traits are adaptations. Some are neutral in terms of adaptive function; some are "free riders" (they came to prevail because some related trait was selected for); and some come about through processes other than natural or sexual selection (Nelson 2003; Sober 1993). As we earlier noted, Sober's claim that "a trait is an adaptation if and only if is selected *for*" (Sober 1993), is a commonly held view among evolutionary theorists. Ironically, this view of adaptation, although relatively straightforward, is difficult to apply. Indeed, the prominent evolutionary biologist G.C. Williams argued that "evolutionary adaptation" is "a special and onerous concept" (G.C. Williams 1966, vii). Williams was in part alluding to cases in which the term "adaptation" is applied to traits for which all that had been established is that they are conducive to fitness – that is, adaptive (Williams 1996, vii). We earlier noted that arguments for a distinction between traits that are adaptations and traits that are only adaptive, date back to Darwin.

This points to a second sense in which adaptation is a difficult notion. To make the case that a trait is an adaptation, we must be able to rule out alternative, equally viable explanations. We earlier noted other forces and possible origins which need to be ruled out, as does the possibility of conversion of function (Nelson 2003). Finally, as Dennett expresses a view we have already noted, when the trait in question is an adaptive behavior or capacity exhibited by a species displaying some degree of behavioral flexibility, equally viable explanations include the discovery of a forced move – a concept borrowed from chess to describe an adjustment in behavior to "stave off disaster" – or a good trick – "a behavioral talent that protects [an organism] or enhances its chances dramatically," and, in the human case, "information transmission and cultural descent" (Dennett 1995, 487–488).

Given such issues, many argue that to make a plausible case that a trait is an adaptation, one needs to make use of history. And this is another reason why adaptation is difficult to establish. The kinds of historical information needed to rule out alternative explanations is often hard to come by – in part, as Sober notes, because natural selection "tends to cover its own tracts ... to destroy the variation on which it acts" (Sober 1993, 69). We have also considered Richardson's arguments about the current inability to use historical methods, such as population genetics and the comparative method, to demonstrate that a human trait is an adaptation.

Reverse Engineering

Earlier we noted that Human Evolutionary Psychologists employ two methods in developing their adaptation explanations and that both make use of reverse engineering. We turn now to issues raised by advocates and critics of using reverse engineering to develop adaptation explanations.

Some maintain that reverse engineering is not only appropriate but indispensable to adaptation explanation. Dennett argues, for example, that

> The work done by natural selection is R and D, so biology is fundamentally akin to engineering, a conclusion that has been deeply resisted out of misplaced fear for what it might imply ... Once we adopt the engineering perspective, the central biological concept of *function* ... can be explained ... The engineering perspective on biology is not merely occasionally useful, not merely a valuable option, but the obligatory organizer of all Darwinian thinking, and the primary source of its power. (Dennett 1995, 185–87; emphasis in original)

Dennett's mention of what he calls "a misplaced fear" refers to the concern raised by critics of the method that beginning from the assumption that a given trait or capacity *is* an adaptation, encourages what Steven Jay Gould and Richard Lewontin call "just so stories" (Gould and Lewontin 1979). Critics of the method, Dennett argues, "were [really only] reacting against a certain sort of laziness: the adaptationist who hits upon a truly nifty explanation for why a particular circumstance should prevail, and then never bothers to test it because it is too good a story, presumably, not to be true" (ibid., 242). As we have noted, evolutionary psychologists do test some of their hypotheses – but the question that warrants attention is whether they consider and/or provide evidentially warranted reasons to reject *alternative* and arguably equally-viable explanations.

However, Dennett also argues that in the human case, "the very considerations that in other parts of the biosphere count *for* an explanation in terms
of natural selection – manifest utility, obvious value, undeniable reasonable-
ness of design – count *against* the need for any such explanation in the case of
human behavior" (ibid., 487–88; emphasis in original). He argues that things
like information transmission must be ruled out to make a plausible case
that a trait or capacity enjoying one or more of these virtues is an adaptation.

Critics of the use of reverse engineering in evolutionary theorizing also
argue that, when it is used by those committed to adaptationism, the
reasoning is often circular. To take an example representative of those
feminists cite, recall Buss' hypothesis that women have an evolved preference for men with resources. This hypothesis, Buss maintains, is derived
from our "knowledge" of an adaptive problem Pleistocene women faced, and
we know of the problem based on studies of contemporary hunter-gatherer
groups. From such knowledge, we can predict a "predisposition" to value
men with resources that was selected for to solve this problem.

The reasoning here, feminists argue, is circular in two respects. For one
thing, the task Buss sets himself is to predict a cognitive mechanism with
which *contemporary* women are endowed on the basis of an adaptive problem
in an *ancestral* population. But, as I have elsewhere argued,

> the adaptive problem alleged to face our female ancestors is derived from
> purported behavior in *contemporary* hunter/gatherer groups. Even if accounts
> of these groups on which Buss and others in the field draw were
> unproblematic (which, we have seen, they are not), these women are not our
> *ancestors*. In addition, the reconstruction of ancestral conditions on which Buss
> relies, *assume* the gender stereotypes (males as providers, females as pregnant
> and dependent, and so forth) from which the adaptive problem is derived.
> (Nelson 2003)

This kind of circularity, feminists argue, is all too common when evolutionary psychologists turn to sex/gender. Donald Symons' defense of a general methodological assumption guiding the discipline may be particularly
telling but it is in no way singular.

> To an evolutionary psychologist, the likelihood that the sexes are
> psychologically identical in domains in which they have recurrently
> confronted different adaptive problems over the long expanse of human
> history is essentially zero. (Symons 1992)

The goal, however, was *to establish* that there *were* sex differences in the adaptive problems faced by our *ancestors* and *to establish* that sex-differentiated, heritable cognitive mechanisms *resulted*. It seems clear, critics argue, that assuming these things from the outset, and failing – in terms of either specific or general hypotheses – to consider alternative explanations for the features of human psychology evolutionary psychologists propose, represents the kinds of circular reasoning to which, critics of adaptationism and reverse engineering charge, they are prone.

Feminist Critiques of Evolutionary Psychology

As we discussed in Chapter 3, the status of sexual selection as an evolutionary mechanism, including questions about whether it is a separate mechanism from natural selection and about whether its role is more than meager, are issues of ongoing investigation and debate in evolutionary biology. We noted that some view sexual selection as Darwin viewed it: an evolutionary mechanism distinct from natural selection (e.g., Dawkins 1989); others view sexual selection as a useful way to think about processes that are, in the end, instances of natural selection (e.g., Bateman 1948); and some maintain that if it is a distinct mechanism, its role is quite small (e.g., Mayr 1972).

We have also discussed feminists' critiques of Parental Investment Theory and how case studies of specific species have provided counter-examples to its predictions. Finally, we noted the emergence in evolutionary biology and in primatology of alternative models of mating and parenting strategies. In short, the theory no longer enjoys the kind of unequivocal acceptance it once did. But, feminists point out, Human Evolutionary Psychologists often do not acknowledge this.

Feminists also criticize specific hypotheses advanced in EP that are based on PIT that we mentioned earlier. Of one of the purported evolutionary explanations of rape considered by Randy Thornhill and Craig T. Palmer, philosopher Elisabeth A. Lloyd criticizes their explanation because she maintains that Thornhill and Palmer do not establish that the cognitive mechanism or predisposition they posit is heritable or that it is not due to factors other than natural selection. Further, she argues, Thornhill and Palmer, among other evolutionary psychologists, incorrectly assume that rape can be approached as a "unitary phenomenon," despite what Lloyd describes as a "striking disunity among the various acts that are classified as rape" (Lloyd

2003). Her argument is representative of those offered by other feminist and non-feminist critics of Thornhill's and Palmers analysis. But, in a review essay, sociologist Sharon Araji also addresses the ways in which Thornhill and Palmer brush aside feminists' and sociologists' critiques as "non-scientific."

> As Darwinists, these theorists see themselves as having the only valid explanation of rape. Throughout the book, they dismiss social science and feminist theories of rape as being nonscientific. (Araji 2000, 2–3)

Despite the fact that many of their feminist critics *are scientists,* this is a common response to feminist critiques by evolutionary psychologists (e.g., Buss 1999). And as we learned in Chapters 2 and 3, there are feminist scientists and science scholars (including this author) who *are* well trained in Darwinian theory *and are* "Darwinians" in some sense, even if some reject sexual selection and most reject Darwin's accounts of sex/gender differences. This, as we will discuss when we turn to relevant philosophical issues, demonstrates that the dichotomy often assumed in EP between science and feminism is not real.

In the case of EP's purported explanations of domestic violence against women and stepchildren, feminists point to what they argue are unwarranted background assumptions on the part of those who propose them. Philosopher Heather Douglas, for example, studies the effects on women of various criminal law systems, and she questions assumptions common in such explanations, including assumptions "that couples are focused on having children; that couples are heterosexual; and that women are primarily attracted to men who will provide". We know, Douglas argues, "that domestic violence goes beyond these categories" (Douglas 2011). As she and others point out, it occurs in same-sex relationships, and in relationships in which partners are not interested in or able to have children. In Chapters 2 and 3, we discussed the critiques feminists have levelled at assumptions of women's and female's dependency; in this chapter, we noted that, although Buss and his colleagues in EP claim that a preference among women for men with resources is confirmed in surveys, this does not establish that the preference is an adaptation. We would need evidence that there were relevant variations in ancestral populations (something that studies of women in *contemporary* hunter-gatherer groups cannot provide) and that the preference is *heritable.* Lacking such information, we would at least need to rule out alternative, and arguably equally viable, explanations for the preference.

Lastly, if we are convinced by Dennett's arguments that a psychological or behavioral feature that is rational is in no need of an evolutionary, adaptation explanation, then Buss' explanation of women's preferring men with resources as being an adaptation can be taken to be wholly unnecessary.

More Philosophical Issues

Evidence, Politics, and Ethics

We have noted in several contexts that EP is viewed by many, including scientists and philosophers not working in the program, to be a serious and, for many, an important extension of evolutionary theory. Its supporters often cite the kind of "conceptual integration" that motivates architects of EP and motivated HS in the 1970s and 1980s. We should also note that practitioners of EP commonly advance an additional line of argument for the conceptual integration of the social/behavioral and biological/natural sciences that they envision and see themselves as bringing about. It starts from the premise that some science is value-laden, of which "old" social science paradigms such as "behaviorism" and "cultural relativism" are taken to be prime examples; and some science is value-free, of which evolutionary biology and cognitive science are examples. Thus, they argue, using evolutionary biology and cognitive science to identify and explain features of human psychology, and behaviors related to them, is the way for the social sciences to achieve the objectivity enjoyed by the natural and biological science.

Critics, including but not limited to feminists, offer two general responses to such arguments. One is that the last five decades have produced compelling evidence, some of which we have considered in this and earlier chapters, that challenges the assumption that science is "value-free." Again, much of this evidence emerged independently of feminists' engagements with science. Moreover, feminists argue that there is important work underway that explores the ways in which non-epistemic values can have a *positive* impact on scientific theorizing (Longino 1996), a topic we discussed in Chapter 4 among other places.

The question of whether evolutionary theorizing is or can be value free has been concretely addressed by defenders and critics of Darwinism. In this chapter, we have explored, for example, feminist criticisms of EP's hypothesis about the evolutionary cause of "domestic violence" that assumes both that domestic partnerships are heterosexual and that the purpose of domestic partnerships is to have children (both assumptions being value laden). A number of other hypotheses about sex/gender advanced in EP are argued by feminists to be informed by gender stereotypes and gendered metaphors. And gender stereotypes and gendered metaphors, they argue, are evaluatively thick – that is, *not* value-free, for reasons considered in previous chapters. Finally, even if one does not view debates about the role of social and political values in good science to be settled, it can no longer be proclaimed or assumed without argument that all science, and in this case, all of evolutionary biology, *is* value-free.

A second response was articulated by Philip Kitcher in *Vaulting Ambition*, in which he offered a sustained critique of Human Sociobiology (Kitcher 1985). Kitcher began by noting that "The dispute about human sociobiology is a dispute about *evidence*." And he maintains that although participants in the dispute might believe that the political implications of sociobiology's explanations should be assessed separately from their evidential warrant, things are not so straightforward. "Although we should accept a hypothesis, whatever its political implications, about human nature *given sufficient evidence*," the question of *how much evidence is sufficient* to warrant the adoption of such hypotheses is not independent of their political implications." (ibid., 9; emphasis in original).

> If a single scientist, or even the whole community of scientists, comes to adopt an incorrect view of the origins of a distant galaxy, an inadequate model of foraging behavior in ants, or a crazy explanation of the extinction of the dinosaurs, then the mistake will not prove tragic. (ibid., 9)

In contrast, Kitcher argues, there may be serious consequences if we adopt an incorrect hypothesis about human nature that carries social and/or political implications. In the case of such hypotheses, research should be held to higher standards; more evidence should be required before such hypothesis are advanced than is required for hypotheses without such implications (ibid., 8–10).

Note that Kitcher is not calling on scientists to abandon a research agenda or a hypothesis because it has social and/or political implications. Rather, he is calling on scientists to uphold the most stringent *epistemic* standards when their hypotheses carry social and/or political consequences. Kitcher's argument also does not call on scientists to wade into the social or political issues at stake – and, indeed, many might not have the background or training to do so, a topic we turn to in the next chapter. Kitcher's claim is that there are ethical responsibilities that come with the cognitive authority scientists are granted and exercise. To meet them only requires doing what one was trained to do: practice science well *and* with due attention to the potential implications of those hypotheses concerning human nature that, as is true of all hypothesis, are *tentative*.

For feminists, the kind of argument Kitcher offers is clearly applicable to the research and hypotheses of EP concerning sex/gender, for much of this research clearly does have social and/or political implications, even though, as we discuss in the next section, its practitioners deny this. Feminists also maintain that the central questions about EP concern *evidence*. Does EP advance hypotheses that are based on uncontroversial aspects of evolutionary biology and/or cognitive science? Are its methods uncontroversial? Are the objects it posits (e.g., "cheater-detection modules" and "a predisposition for rape") scientifically respectable (e.g., do they have clear identity criteria)? Do EP's hypotheses represent the only viable explanations of the phenomena in question, and do its advocates consider alternative explanations? Does EP test its hypotheses and, if so, are apparent confirmations taken to rule out the need to consider alternative explanations of the phenomenon at issue?

Finally, if we view EP (as some of its defenders do) as "*a* promising *new* research program" (setting aside its reliance on PIT and its parallels with HS), do its practitioners – as some would argue even those of well-established programs should – admit to and be clear about the tentativeness of their assumptions, claims, and hypotheses? Given that their hypotheses about phenomena such as rape and domestic violence can be understood to have social and ethical implications, are caveats to the effect that to claim a feature of human psychology is an evolutionary adaptation that may now be *maladaptive* sufficient to blunt any argument that those who propose such features have an ethical obligation to uphold the most stringent standards of evidence?

Feminist Critiques and the Naturalistic Fallacy

We conclude with arguments offered by evolutionary psychologists that the hypotheses advanced in the discipline do not carry normative implications;

and that critics, particularly feminists, who ascribe such implications to them, commit the naturalistic fallacy. We begin with a discussion of "the naturalistic fallacy."

The Naturalistic Fallacy

The British empiricist David Hume (1711–1776) is generally credited with first identifying the pattern of reasoning that has become known as "the naturalistic fallacy" as being fallacious (i.e., flawed) reasoning, although Hume did not use that phrase. The philosopher G.E. Moore is generally credited with introducing the phrase "naturalistic fallacy." We focus on Hume's argument about the naturalistic fallacy because this is the argument evolutionary psychologists emphasize. We do not consider whether Hume was correct in claiming that the reasoning in question is fallacious. This is an important question but it is not relevant to the arguments we do consider, arguments that are important in their own right.

Hume drew a distinction between two kinds of statements: statements about "matters of fact" and statements concerned with what he called "relations of ideas." Matter of fact statements are what we would call empirical or factual statements. "It is below freezing today in Minnesota" is a statement of this sort. Whether it is true is depends on what the weather in Minnesota is like today. Statements about relations of ideas, on the other hand, are not about what the world is like, but rather about concepts. Their truth or falsity depends on whether the concepts that they relate are or are not so related. They are generally taken to include what we call logical truths, for example, "A square has four sides," as well as logical falsehoods, such as "Some bachelors are married."

Hume notes that he has observed cases in which authors advance one or more empirical claims (i.e., "matter of fact claims"), and then without comment proceed to advance a normative conclusion – that something *ought* to be the case.

> In every system of morality, which I have hitherto met with, I have always remarked that the author proceeds for some time in the ordinary way of reasoning [noting or establishing is, or is not, statements] . . . when of a sudden I am surpriz'd to find that instead of the usual copulations of predicates *is*, and *is not*, I meet with no proposition that is not connected to an *ought*, or an *ought not*. (Hume 1739, 335 [Hume 1965, 469]; emphasis in original)

By "I meet with no proposition that is not connected to an *ought*, or an *ought not*," Hume means that the propositions [statements] now before him are all "ought" or "ought not" statements. "This change," Hume continues, "is imperceptible [not noted or remarked on by the author]; but is, however, of the last [i.e. greatest] consequence." Such reasoning, Hume argued, is fallacious: "ought" or "ought not" statements *do not follow from* "is" or "is not" statements. No claim about what is or is not morally right, Hume maintained, can be established by appealing only to empirical claims.

Evolutionary psychologists argue that they do not in fact commit the naturalistic fallacy. They argue that the specific features of human psychology and evolutionary explanations of them that they propose are purely factual – that is, *is* claims. Because they agree with Hume that to draw normative conclusions from such claims would be fallacious, most avoid doing so. But, evolutionary psychologists argue, feminist scientists and science scholars who criticize their hypotheses about sex/gender, *do* engage in reasoning of the kind Hume maintained is fallacious. They take feminists who criticize their hypotheses as drawing normative implications (i.e., social, political, and/or ethical implications) from the hypotheses about sex/gender they propose, implications that, according to evolutionary psychologists, are not implied by them. The third and last argument we consider relevant to these issues questions whether the hypotheses EP proposes about sex/gender *are* purely factual or empirical. Feminists' critiques of these hypotheses contend that the hypotheses in question are "evaluatively thick," that they carry normative as well as factual content and thus are not purely "is" claims.

Books and textbooks in EP often include a description of the naturalistic fallacy that is based on Hume's argument, and maintain that the hypotheses proposed or discussed in the volume do not commit that fallacy – i.e., do not entail normative conclusions. For example, Steven Gaulin and Donald McBurney, authors of the textbook, *Psychology: An Evolutionary Approach*, state

> Evolutionary psychology explains behavior; it does not justify it. Imagining that it offers a justification is known as the naturalistic fallacy (Buss 1990). In a nutshell, the naturalistic fallacy confuses "is" with "ought". It confuses the situation that exists in the world with our ethical judgement about that situation. (Gaulin and McBurney 2000, 2)

We earlier noted that in *A Natural History of Rape: Biological Bases of Sexual Coercion*, biologist Randy Thornhill and anthropologist Craig T. Palmer argue that rape is either an adaptation (i.e., the product of natural or sexual selection) or a byproduct of other adaptations (for example, sexual desire); and they insist that these explanations do not entail that it is justified. Rather, Thornhill and Palmer argue, knowing the biological bases of rape can lead to preventative measures – for example, to education programs for teenagers devoted to rape prevention (Thornhill and Palmer 2000, 179).

Evolutionary psychologists' arguments that a cognitive predisposition for behavior that was once adaptive may be "maladaptive" now (for example, a disposition among men towards sexual jealousy) are understood to demonstrate that evolutionary psychologists do not draw normative conclusions from the purely factual hypotheses they claim to advance. But what does it mean for a trait, including a psychological trait, to be *maladaptive*? To describe an earlier adaptation as now maladaptive would seem to imply that the trait in question now works against, rather than for, survival or reproductive success. But the EP literature makes it clear that this is not generally what evolutionary psychologists mean when they describe an earlier adaptation as now maladaptive. Consider, for example, Steven Pinker's argument that our fear of snakes, adaptive in ancestral environments, is now maladaptive given that today most humans live in urban environments. In this and similar cases (e.g., in discussions of our fear of heights), "maladaptive" can only mean "irrational" or "unwarranted." There is no reason to think, nor does Pinker argue, that such fears work against our survival or reproductive success (Pinker 2009).

But then why do evolutionary psychologists use the term "maladaptive"? Do arguments that describe child abuse or rape as "maladaptive now" – taken to demonstrate that EP does not conflate "is claims" with "ought claims" – assume that the behavior in question is *unethical*? If so, it would seem that, in describing a trait as "maladaptive," evolutionary psychologists do advance normative claims. Now there is a way to avoid this conclusion although evolutionary psychologists do not seem to make use of it. Within the framework of Darwinism, a predisposition in relation to a behavior or the behavior that results is maladaptive if and only if it acts against the survival or reproductive success of *an individual so predisposed*. Given that individuals who engage in child abuse or rape are subject to imprisonment or even death in some cultures, such behavior is not conductive to their survival or reproductive success, and thus it is maladaptive. This interpretation is certainly in keeping with how "maladaptive" is understood in evolutionary theory. But, as far as I know, no evolutionary psychologist has made use of it. So, we are left wondering whether, in fact, the reasoning that leads them to call some predispositions or behaviors "maladaptive" does not include implicit ethical premises, despite their claims that the hypotheses proposing such predispositions are purely factual.

As we have noted, some evolutionary psychologists charge feminists with committing the naturalistic fallacy by mistakenly attributing normative implications (i.e., political, social, and/or ethical implications) to the empirical hypotheses about sex/gender advanced in the program (e.g., Buss and Schmitt 2011). Another argument taken to demonstrate that feminist critiques of EP commit the naturalistic fallacy maintains that the goals of evolutionary psychology and of feminists are, by their very nature, opposed to one another. EP, proponents of this argument maintain, is like any science in that it seeks to discover "how things *are*." In contrast, the argument continues, feminism is a *political* commitment focused on "how things *should be*" and, as such, is *not* science. In the first edition of the textbook, *Evolutionary Psychology: The New Science of the Mind*, David M. Buss offered this argument (Buss 1999).

So, one important question to ask is this. Are the hypotheses about sex/ gender evolutionary psychologists advance *purely factual*? We have considered feminists' arguments about the presence, function, and consequences of gender stereotypes that are appealed to or relied on in EP. For example, as we have noted, feminists point to the hypothesis that women seek mates who are likely to provide for them and their offspring, as invoking a gender stereotype. Not all women seek or need such support.

We have considered other hypotheses that feminists argue assume or rely on gender stereotypes, which carry normative content because they rule out the possibility of exceptions. And we have seen that feminists maintain that gender stereotypes have led to androcentric reconstructions of the adaptive problems our ancestors faced in the Pleistocene – reconstructions, feminists argue, that are *necessary to* EP's hypotheses about gender-specific predispositions. For these reasons and others considered in previous chapters, feminists maintain that gender stereotypes are evaluatively thick – they include normative as well as empirical content.

Feminists also point to the presence and role of "gendered metaphors" in EP, including its appeals to Parental Investment Theory, which describes eggs as "expensive" and sperm as "relatively cheap." These metaphors, which serve to motivate hypotheses about differences in women's and men's mating and parenting strategies, are *not* literal, empirical descriptions. "Expensive" and "cheap" carry normative content, as do claims about sex/gender differences in "parental *investment*" (e.g., Tang-Martinez 2000). As we saw in our earlier discussion of feminist critiques of specific hypotheses about sex/gender proposed in EP, feminists point to many more examples of evaluatively thick concepts and hypotheses in the research program.

If feminists are correct, then it is arguable that their critiques of EP's hypotheses about sex/gender do not commit the naturalistic fallacy. The hypotheses whose normative implications they criticize are *not* purely factual or empirical. Nor, it follows from this line of argument, do evolutionary psychologists who advance hypotheses about sex/gender commit the fallacy – albeit, *for a different reason*. It is not because evolutionary psychologists avoid drawing normative conclusions from purely factual hypotheses. It is because the hypotheses in question are *not* purely factual *and* the assumptions taken to support them are not either.

This brief discussion makes it clear that the issues involved in clarifying the relationships between the naturalistic fallacy, evolutionary psychologists' hypotheses about sex/gender, and feminist critiques thereof are complex. We should also note that discussions and analyses about evolutionary psychology and feminism, including the kinds of disagreement we have considered, continue to evolve. Contributors to two relatively recent collections of essays discuss them in depth (Buss and Malamuth 1996; and Gowaty 1997).

In the next chapter, we continue our consideration of what, if any, relationships obtain between science and ethics, and the importance feminists, and other scientists and philosophers, have come to attribute to this issue.

9　Socially Responsible Science and Socially Relevant Philosophy of Science

Science and Ethics

We have seen that feminist scientists and science scholars understand their engagements with biology to involve not only epistemic issues, but also important ethical issues. We have considered feminists' arguments about the implications of hypotheses proposing innate sex/gender differences, including how such hypotheses can impact individuals' sense of their own abilities or lack thereof; shape public perceptions about differences in the abilities, preferences, and temperament of girls and boys, and of women and men; and influence educational and social policies. Clearly issues such as these are ethical issues.

Feminists also argue, contrary to some of their critics, that the cases on which they focus, including those we have considered, cannot reasonably be written off as "bad science." Were the cases to be so viewed, one could reasonably conclude that they provide little if any actual insights into scientific reasoning or interrelationships between science and the sociocultural contexts in which it is undertaken. But we have seen that feminists argue that each of the hypotheses we have considered is, or was, when it was proposed, at least fundamentally compatible with, if not a consequence of, the general empirical assumptions and research priorities of the research program in which it is or was proposed. In this respect, these hypotheses are representative of many others feminists have engaged. Hence, feminists argue, the ethical issues in question are raised by *good science*, and thus warrant the attention of scientists and science scholars.

Nor, feminists argue, again contrary to the claims made by some of their critics, are their engagements with biology, both critical and constructive, motivated by a desire to make science "politically correct." The feminist arguments we have considered reflect a simple, and upon reflection surely obvious, concern, that hypotheses about sex or sex/gender, and the assumptions underlying and informing them, be empirically warranted. As feminists understand it, the other major goal of their engagements with biology is to promote ethically responsible science: encouraging scientists to be transparent about the limits, as well as the perceived strengths, of the evidence they cite as supporting their hypotheses; to balance such perceived limits with the potential consequences of their hypotheses for public perceptions and policies and to undertake efforts to identify and critically evaluate any untested and/or evaluatively thick assumptions that inform their reasoning.

We will see that feminists' concerns about relationships between science and ethics are not unique. There is growing interest among scientists representing a variety of disciplines, professional science associations, funding agencies such as The National Science Foundation and National Institutes of Health, public policy makers, philosophers of science, and other science scholars, in identifying factors that can contribute to "socially responsible science."

For example, in *The Liberal Art of Science*, the American Association for the Advancement of Science (AAAS) argues that the teaching of science, at both the graduate and undergraduate level, should include curricula that result in students coming to understand "how scientific knowledge influences and is influenced by the intellectual traditions of the culture in which that knowledge is embedded" (AAAS 1990, 29), and "the interplay between science and the intellectual and cultural traditions in which it is firmly embedded" (ibid., 14). And in a section entitled "The Ethical, Social, Economic, and Political Dimensions of Science," the authors argue that

> Liberal education in the sciences must provide students with linkages to the real world by exploring the values inherent in science and technology, by examining the institutions that set directions for science and technology, and by stressing the choices scientists, citizens, and governments make about science in human lives. (ibid., 14)

One response to the arguments so far summarized is that it is inappropriate to ask or demand that scientists, *qua* scientists, consider not just the empirical warrant for their hypothesis but also the ethical implications of these hypotheses. Attention to this issue and others that concern relationships between science and ethics are not new, and the issues involved are not simple. They involve philosophical assumptions on the part of scientists, philosophers of science, and other science scholars, about the very nature of science. For example, if one takes the fundamental goal of science to be attaining knowledge, and views knowledge to be "a good in itself," one might reasonably conclude that scientists' responsibilities are purely epistemic. From this perspective, one might believe that a responsibility to deal with any ethical and/or social issues attendant to scientific knowledge or technologies is the responsibility of society at large, or of scholars who study the impact of science on society, but surely not the responsibility of scientists whose only job is to produce knowledge.

In the debates following World War II about whether the development of the atomic bomb was ethical, Robert Oppenheimer, a major figure in its development, defended the project by appeal to assumptions very like these. Our purposes here do not include considering the national or military rationale for the atomic bomb, which included information that Germany was engaged in research to build one. Nor is it to consider the ethical debates about its development by America. We focus on Oppenheimer's views about the epistemic responsibilities of scientists and about whether they also have ethical responsibilities in acquiring knowledge, as versions of them have been accepted by many scientists and to a certain extent by society more generally. In response to critiques of the development of the bomb, Oppenheimer argued that

> The reason we did this job is that it was an organic necessity. If you are a scientist you believe it is good to find out how the world works, that it is good to turn over to mankind at large the greatest possible power to control the world, and to deal with it according to its own lights and values. (quoted in Kunetka 1982, 120)

Oppenheimer's view of scientists' ethical responsibilities can be interpreted in two ways. He may be arguing that any ethical issues raised by scientific knowledge or technology are to be dealt with by citizens, which his comments might be understood to suggest does not include scientists. Alternatively, Oppenheimer may be understood as arguing that scientists have no "special" ethical responsibilities beyond those of other citizens.

Whichever is the case, it is important to remember that nonscientists could not have developed the atomic bomb. Nor could nonscientists develop the hypotheses about the specific biological origins of purported sex/gender differences that we have considered. Each required substantial theoretical knowledge. And considering the debates in the United States about climate change and evolution, one can also reasonably ask whether many citizens have sufficient knowledge about science or about ethics to contribute in meaningful ways to deliberations about the ethical implications of scientific research. Given these realities, together with scientific fallibility, one could reasonably conclude that scientists do have a responsibility to consider the ethical issues raised by and possible consequences of the hypotheses and technologies they develop.

Yet, in describing the development of the atomic bomb as "an organic necessity," Oppenheimer suggests that if a scientist does not pursue knowledge that it is possible to obtain, she or he has ceased *to be* a scientist. And this would seem to rule out the possibility that scientists can or should allow ethical considerations to inform their decisions about whether to pursue specific kinds of knowledge or technologies. The only responsibility Oppenheimer attributes to scientists that might be considered "non-epistemic" is that of providing humanity with knowledge that will contribute to its ability, in his words, "to control the world."

Other episodes in the history of science and in contemporary science illustrate that the question of whether scientists have ethical as well as epistemic responsibilities is not new. Many viewed Galileo's and Darwin's hypotheses as challenging long-held beliefs and values, and as undermining accepted ethical systems. Developments in twentieth and twenty-first century science have also led to questions about the ethical issues raised by specific research projects and technologies. There have been debates about the ethics of the development and use of recombinant DNA techniques, and of genetic testing and genetic engineering (including, most recently, genetic editing), to name just a few. And currently there is a debate about what some take to be ethical issues raised by what has come to be called "race-based medicine."

In some cases, those questioning the ethics of a research project or technology have questioned the longstanding assumption that knowledge *is* a good in itself. And in some cases, critics of research projects or technologies emphasize ethical issues raised by human, including scientific, fallibility. A debate in the 1970s about recombinant DNA technology serves as an example in which critics of the technology raised both issues. One thing that makes the debate relevant to our discussion is that those engaged in it were in fact scientists whose views stand in stark contrast to Oppenheimer's arguments that, for scientists, pursuing knowledge that can be pursued is an "organic necessity." As much to the point, arguably the ethical issues that critics raised about using recombinant DNA technology, parallel the ethical concerns feminists raise about the hypotheses about sex/gender differences that we have considered.

Recombinant DNA technology allows scientists to cut and splice segments of DNA and transfer genes from one organism to another, including from a member of one species to that of another. This allows scientists to override barriers nature has built between species, and to create new forms of life. When it was developed in the early 1970s, the technology was recognized, in the words of science commentator Nicholas Wade, as "a major turning point ... in the study of life, a turning point of such consequence that it may make its mark not just in the history of science, but perhaps even in the evolution of life itself" (Wade 1977, 3).

Concerns about unpredictable short-term and long-term consequences of using the technique (respectively, the creation of pathogens, and consequences for the directions of evolution) led some prominent biologists to call for a voluntary and world-wide moratorium on its use – despite its potential benefits (for example, it allows scientists to create insulin). The call came in the form of a letter signed by some members of the National Academy of Sciences, including Nobel Laureates such as James Watson, published in both *Nature* and *Science*. It at least appeared to be successful and, as far as anyone knew, research using the technique did not resume until the NIH developed strict guidelines for its use that many biologists took to address the concerns that prompted the moratorium. (Of course, even during the moratorium, there was no guarantee that all scientists or all countries *would* observe it.)

We focus on the ethical concerns that continued to be voiced by scientists who favored a permanent moratorium. Molecular biologist Erwin Chargaff's letter "On the Dangers of Genetic Meddling," which was published in *Science*, was representative in voicing the concerns of some biologists. In it, Chargaff questioned whether, given human fallibility, the limits on our abilities to

control nature, and the potential consequences of the technology, recombinant DNA technology can be justified.

> What seems to have been disregarded completely is that we are dealing here
> much more with an ethical problem than with one in public health, and that
> the principal question to be answered is whether we have the right to put an
> additional fearful load on generations that are not yet born.
>
> You can stop splitting the atom, you can stop visiting he moon, you can stop
> using aerosols, you may even decide not to kill entire populations by the use of
> a few bombs. But you cannot recall a new form of life. Once you have
> constructed a viable E coli cell carrying a plasmid DNA into which a piece of
> Eukaryotic DNA has been applied, it will survive you and your children and
> your children's children.
>
> Have we the right to counteract irreversibly the evolutionary wisdom of
> millions of years, in order to satisfy the ambition and curiosity of a few
> scientists?

"An irreversible attack on the biosphere," Chargaff concluded, "is something
so unheard of, so unthinkable to previous generations, that I could only wish
that mine had not been guilty of it" (quoted in Wade 1977, 104–105).

Robert Sinsheimer, another prominent biologist who advocated a permanent moratorium, challenged the view that any knowledge that can be
attained, should be. Citing human fallibility, Sinsheimer argued that,
although "the credo 'Know the truth and the truth shall make you free' is
carved on the walls of laboratories and libraries across the land,"

> We begin to see that the truth is not enough, that the truth is necessary but
> not sufficient, that scientific inquiry, the revealer of truth, needs to be coupled
> with wisdom if our object is to advance the human condition ... Armed with
> the powers [now provided us by the physical and biological sciences], I think
> there are limits to the extent to which we can continue to rely upon the
> resilience of nature or of social institutions to protect us from our own follies
> and finite wisdom. Our thrusts of inquiry should not too far exceed our
> perception of their consequences. (quoted in Wade 1977, 107)

Whether one agrees with Chargaff and Sinsheimer, they serve as examples of
scientists who believe that scientists have ethical as well as epistemic
responsibilities.

And there are some obvious parallels between the ethical concerns the
two raised, and the epistemic and ethical issues feminists argue are raised by

scientific research and hypotheses about sex/gender differences that we have considered in this text. Feminists also stress the epistemic issue of scientific fallibility and the potential ethical consequences of claims and hypotheses such as:

- Given the lesser amount of pre- and postnatal testosterone to which human females are exposed, they are unable to engage in math and science as successfully as human males.
- For the same reason, women are less likely to exhibit the types of aggression some scientists have been taken to be necessary to achieving leadership and power.
- The selection pressures to provide and protect women and offspring, and to protect territory that men face (or at least did in the Pleistocene), has resulted in their superior reasoning and observational abilities relative to women.
- Given gametic dimorphism and women's greater parental investment it is, as David Barash made the point, "... perfectly good biology that business and profession taste sweeter to [men], while home and child care taste sweeter to women." And his claim that, women "seeking liberation" by working outside the home, may fall into "a socially instituted trap that is harmful to everyone concerned" (Barash 1979, 114–115).
- Nineteenth-century medical hypotheses that women's reproductive organs are the source of all their illnesses and diseases, and the removal of women's ovaries and uteruses to treat them.
- Twentieth and twenty-first century medical hypotheses that many if not most women suffer from PMS, that define menopause as "estrogen-deficiency," and what many argue are unnecessary interventions to "treat" these "conditions."

These are just a few of the hypotheses we have considered, as well as arguments offered by feminist biologists that they are or were actually or potentially consequential for public perceptions of sex/gender differences and appropriate gender roles. In terms of their impact on girls and women, they are not unlike the hypothesis proposing women's intellectual inferiority offered by nineteenth century anthropologists and psychologists based on the smaller size of women's brains mentioned in Chapter 3. As a then contemporary scientist who rejected the hypothesis of women's intellectual inferiority noted about the "measurements" purportedly showing women to have smaller brains than men,

Women displayed their talents and their diplomas. They also invoked philosophical authorities [such as John Stuart Mill]. [But] these numbers fell on women like a sledge hammer ... Theologians had asked if women had a soul. Several centuries later, some scientists were ready to refuse them a human intelligence. (L. Manouvrier, quoted in Gould 1980)

So, there are parallels between the ethical and epistemic concerns raised by critics of recombinant DNA technology, on the one hand, and the ethical and epistemic concerns feminists raise about hypotheses proposing sex/gender differences, on the other. Both emphasize the fallibility of science and the social implications of areas of scientific research. But there is also a difference. Most feminists do not argue that research on sex differences should never be undertaken. They argue that when it is appropriate to undertake such research, it should be done carefully and responsibly.

As noted earlier, there are reasons why disagreements remain about the relationships between science and ethics. Some, such as Oppenheimer, view scientists' responsibilities as purely epistemic. It is also the case that many contemporary scientists have no training or background in ethical theory and, therefore, one can reasonably ask if they are in a position to engage ethical issues without collaborating with others who are so trained. Another reason for such disagreements, which we have had reason to note in several contexts, is the longstanding distinction drawn between "facts" and (non-epistemic) "values" – a distinction that in part reflects the view among many empiricists that, unlike empirical claims about "facts," value claims, including ethical claims, are not subject to empirical constraints. Thus, for generations of scientists and philosophers, non-epistemic values have been viewed as inherently compromising of scientific inquiry – and science, at least "good science," has been viewed as free of them. The very idea that a scientist should allow potential ethical issues to override her or his efforts to acquire knowledge about nature seems to some to raise the question of whether scientific objectivity would be compromised (e.g., Resnik and Elliott 2016).

Such concerns are important. But as we next explore, some scientists and philosophers are seeking to address them, as their perspectives about the kinds of relationship that do or should obtain between science and ethics are changing.

Socially Responsible Science

As reflected in recent articles, books, and conferences, there is growing interest among scientists, philosophers of science, and science scholars in exploring scientists' "social responsibilities." As the authors of a 2016 article reviewing relevant publications and conferences note,

> Numerous scientists and philosophers have argued that scientists have a responsibility to address the social implications of their research ... Many professional codes specifically mention duties related to social responsibility in science (e.g., American Anthropological Association, American Chemical Society, [and] American Society for Microbiology). The National Institutes of Health (NIH) requires that funded students and trainees receive instruction in the responsible conduct of research (National Institutes of Health 2009). (Resnick and Elliott 2016, 31)

Resnick and Elliott also note that the National Science Foundation (NSF) now requires grant proposals "to address the social impact of research [for which funding is requested]" (Shamoo and Resnick 2014). Even a cursory look at the issues addressed in the books and articles Resnick and Elliott review about the "social implications" and the "social impact" of scientific research makes it clear that many of these issues are ethical issues. At the same time, Resnick and Elliott argue that "recognizing one's social responsibilities as a scientist may be an important first step toward exercising social responsibility, but it is only the beginning, since scientists may confront difficult value questions when deciding how to act responsibly" (ibid., 31). Noting the differences in the priorities and issues raised in the relevant literature, and the balance they suggest needs to be maintained between one's professional and social responsibilities, which they take to be somewhat different, they conclude that the ethical issues and dilemmas related to socially responsible science warrant much more investigation.

The American Association for the Advancement of Science (AAAS) is also addressing the general issue, noting that "The notion that scientists have a responsibility to society that goes beyond their responsibilities to the profession is longstanding." Like Resnick's and Elliott's review, AAAS makes it clear in a 2015 statement posted on their website that the literature devoted to the topic indicates disagreements about just what those responsibilities are.

> There is no consensus on what the content and scope of [scientists'] social responsibilities are or ought to be. While there is a growing literature

concerning the issues encapsulated by the phrase "social responsibilities of scientists," a review of the literature reveals many and sometimes competing views, and the lack of data to inform the discussion. (AAAS 2015)

In an effort to get clearer about areas of agreements and disagreements about how scientists view their social responsibilities, AAAS developed and distributed an online questionnaire to scientists, engineers, and health professionals internationally. The goal was to use the data provided by the survey to create a second survey that they hope will yield a "more representative and scientifically rigorous" picture of the perspectives of the international community of scientists. They published the results of the preliminary survey in *Social Responsibility: A Preliminary Inquiry into the Perspectives of Scientists, Engineers, and Health Professionals* (Wyndham et al. 2015).

In the executive summary, the authors provide a list of the 10 "social responsibilities" respondents were asked to rank on a five-point scale ranging from "Critically important" to "Not at all important," as well as an option to choose "unsure," and it provides a summary of the responses. There was strong consensus (between 92 and 95 percent) that scientists should "take steps to minimize anticipated risk" and "to consider the risks of adverse consequences associated with their work." There was somewhat less consensus (82.4 percent) that scientists "should take steps so that their research, findings or products are not used inappropriately by others." The summary reported other differences in perspectives; for example, younger respondents took it to be more important to "explain their work to the public" than did older respondents, for whom "reporting suspected misconduct" was ranked as more important. And, as the literature review that prompted the survey suggested, the summary reported differences across disciplines in how important responders took a specific responsibility to be, with engineers, for example, finding a responsibility to explain their research to the public much less important than did health care professionals. And the survey also revealed global regional differences in the importance attributed to specific responsibilities.

Based on these differences, the authors cited the following as among the most important questions to be addressed in the follow-up survey: "What factors (institutional structures, domestic legal and ethical frameworks, disciplinary codes of conduct) influence individuals' perceptions of their social responsibilities?" and "If scientists draw a connection between their professional and social responsibilities, what kinds of public commitments do they recognize, and how do they establish priorities (if at all) among these priorities?" These are the kinds of information they maintain that it is important to acquire to identify ways to foster socially responsible science.

So, on the one hand, there is obviously interest among scientists representing a range of fields and cultural contexts in the issue of socially responsible science. On the other hand, there are disagreements, as well as uncertainties, about just what those responsibilities are, which are most and least important, and how they can be fulfilled. On the other hand, we have noted publications and studies that indicate that attention to issues involving the social responsibilities of science will continue in the future.

Yet it is also important to notice that in terms of the disagreement about whether scientists have ethical as well as epistemic responsibilities as reflected in the views so far considered, there seems to be a general assumption that the issues in question only pertain to scientists or, among those who hold views in keeping with Oppenheimer's, only pertain to nonscientists, in the exclusive sense of "or." And the assumption that seems most widely accepted is that bringing about socially responsible science is solely the responsibility of scientists. But is this assumption correct? We next turn to different approaches to the issues of socially responsible science and to relationships between science and ethics. They assume that responsibility for studying and reaching decisions about the ethical and/or social impacts of scientific research should be broadly shared, by policy makers, ethicists, science scholars knowledgeable about one or more sciences, ethicists, and social scientists, among others, as well as scientists engaged in the physical and biological sciences.

Socially Relevant Philosophy of Science

Interest among some philosophers of science in making their discipline "socially relevant" led to a conference devoted to the topic during the 2008 meeting of the Pacific Philosophy of Science Association, and to a special issue of the journal *Synthese* (*Synthese* 2010) that included papers presented at the conference as well as others. Many of those who spoke at the conference, and the editors of the special issue of *Synthese* as well as contributors to it, are feminist scientists and philosophers of science. As will become clear, there is a strong connection between interest in socially responsible science and in socially relevant philosophy of science. The core aspiration of those interested in making philosophy of science socially relevant is to find ways to foster and promote socially responsible science. The organizers, presenters, and participants in the conference shared the view that work undertaken in philosophy of science could and should be expanded to encompass more than the emphases that dominated the field in the mid-to-late twentieth century – namely, the analysis of scientific reasoning.

Although conference presentations and papers in *Synthese* often noted that work in the philosophy of science had already undergone significant changes in that more philosophers are now engaging specific sciences and research programs, and more are working in collaboration with scientists, they called on those in the discipline to further expand the issues they engaged and the activities they undertook. As the introduction to the special issue of *Synthese* described the goals of the conference and articles, they were to explore ways in which philosophy of science "can provide social benefits, as well as benefits to scientific practice and philosophy itself" (Fehr and Plaisance 2010, 301).

That "expansion" of the discipline's focuses was envisioned, rather than abandonment of some of the discipline's emphases, reflected assumptions that in investigating how the discipline could be socially relevant, rather than a relatively insular enterprise, analyses of scientific reasoning – including methodological, epistemological, and metaphysical commitments – would remain important to this work. The difference envisioned involves how philosophers can expand such analyses in ways that are relevant to the impact of scientific research on the lay public, or subgroups thereof, and in ways that can contribute to scientists' abilities to positively engage with members of the various publics affected by their research. In these respects, the editors of the journal issue argued, socially relevant philosophy of science is fundamentally pluralistic; "it includes philosophical engagement with scientific research on socially relevant topics, [and] philosophical activities that attend to the interactions among scientists and various communities that contribute to and are affected by scientific research" (ibid., 302). The editors also pointed out that there was precedence for focusing on socially relevant issues. The first by-laws for the PSA (Philosophy of Science Association), as laid out in 1948, stated that the new society was dedicated to "a furthering of the study and discussion of the subject of the philosophy of science, broadly interpreted, and the encouragement of practical consequences which may flow therefrom of benefit to scientists and philosophers in particular, and to men of good will in general" (ibid., 311).

Although changes in philosophy of science were emphasized, the conference presentations and articles in the special issue of *Synthese* did reflect the interdisciplinary nature of the work envisioned. They included analyses of the impact of specific areas of scientific research on what they called "various stakeholder groups," including policy makers, various publics, and scientific practitioners. One contributor, philosopher Heather Douglas, provided an analysis illustrating that engaging with scientists "on the ground" facilitated her development of normative guidelines for scientists involved in risk analysis. Contributors also emphasized the need to create venues for interactions between scientists and those whom their research will impact, seeking to foster trust relationships, knowledge among the relevant public of the rationale and assumptions of the research in question, and knowledge on the part of scientists of the perspectives of such publics about potential impacts of that research. The "publics" contributors took to be relevant to these projects are themselves diverse, including marginalized groups, indigenous groups, and the disabled, for whom the social impact of scientific knowledge and technologies might well be different from its impact on others.

The articles in the issue of *Synthese* also identify obstacles (some institutional, some based on philosophical assumptions) to bringing about the kinds and degree of collaborative efforts undertaken by scientists, lay publics, and philosophers that their authors believe can benefit members of all three groups. They also offer concrete suggestions for overcoming such obstacles and fostering such collaborations.

Socially Relevant Philosophy of/in Science and Engineering

A more recent development related to interest in "socially responsible science" is the founding of The Consortium for Socially Relevant Philosophy of/in Science and Engineering (SRPoiSE). Some of those instrumental in its formation, including Heather Douglas, Carla Fehr, and Kate Plaisance (all of whom engage in feminist philosophy of science), were also instrumental in fostering interest in "socially relevant philosophy of science." The addition of "in" in the name of the consortium signals a strong interest in directly engaging with scientists and engineers to encourage them to also engage issues related to their research that are socially relevant. The mission statement of SRPoiSE parallels but also expands upon the goals and projects envisioned in earlier discussions of socially relevant philosophy of science in the emphasis it places on the need for interdisciplinary collaborations in relation to fostering socially responsible science.

> This consortium supports, advances, and conducts philosophical work that is related to science and engineering and that contributes to public welfare and collective wellbeing. We aim to improve the capacities of philosophers of all specializations *to collaborate and engage with scientists, engineers, policy makers, and a wide range of publics* to foster epistemically and ethically responsible scientific and technological research.
>
> We are particularly interested in addressing *complex social and environmental problems and in fostering the ability of researchers in science and engineering to do so as well. We seek to understand and ameliorate conceptual and institutional barriers to collaborative research across these groups.* (SrPoiSE, emphasis added)

As its mission statement makes clear, the consortium views collaborations among scientists, philosophers, and members of many other groups as necessary to achieving epistemically and ethically responsible scientific research.

The consortium has hosted conferences attended by scientists, engineers, philosophers, and members of other academic disciplines. And its members have undertaken collaborations with scientists. We earlier noted Heather Douglas' collaborations with scientists to develop normative guidelines for risk analysis. Philosopher Nancy Tuana engaged in collaborative projects with climate scientists, and national and international climate policy makers, to address issues involving the nature and social impact of policy, regulation, and institutional structures (Tuana 2010).

These are just two of the projects undertaken by those who want to see philosophy of science become more socially relevant. They are representative in terms of the emphasis they place on encouraging scientists and others to work collaboratively to identify the social impact and social implications of scientific research and technologies, and thus foster socially responsible science.

Conclusion

We have noted that there are disagreements and important open questions about the social responsibilities of scientists that scientists, philosophers of science, and ethicists are currently pursuing. Yet it is also clear that interest in these important issues is strong and that it is growing.

It is also clear that feminists' interests in relationships between ethics and science align with those of many other scientists and philosophers. Common concerns include issues such as inductive risk, values in science, scientific fallibility, and the consequences of scientific research for human well-being and public policy. And although it is not possible to predict what directions scientific and philosophical attention to fostering socially responsible science will take in the future, clearly other scientists and science scholars share feminists' interests in fostering such science. Hence, it seems likely that scientists and science scholars, both feminists and non-feminists, will find ways to work collaboratively in their efforts to bring about socially responsible science. Such collaborations, as we have seen, are already occurring through the auspices of The Consortium for Socially Relevant Philosophy of/ in Science and Engineering.

Bibliography

Altmann, J. 1974. Observational study of behavior: sampling methods, *Behaviour* 49.
1980. *Baboon Mothers and Infants*. Chicago: University of Chicago Press.

American Association for the Advancement of Science (AAAS). 1990. *The Liberal Art of Science: Agenda for Action: The Report of the Project of Liberal Education and the Sciences*. Washington, DC: AAAS.

Anderson, E. 2004. Uses of value judgments in science: A general argument with lessons from a case study of feminist research on divorce. In *Hypatia Special Issue: Feminist Science Studies*, eds. L. H. Nelson and A. Wylie. Bloomington: Indiana University Press.

2015. Feminist epistemology and philosophy of science, in *The Stanford Encyclopedia of Philosophy* (Fall 2015 edition), ed. Edward N. Zalta, URL: http://plato.standford.edu/archives/fall2015/entries/feminism-epistemology/.

Araji, S. 2000. Review essay – *A Natural History of Rape: Biological Bases of Sexual Coercion*, *Alaska Justice Forum* 17(2): 2–3.

Barash, D. 1979. *The Whisperings Within*. New York: Harper & Row.

Barkow, J. H., L. Cosmides, and J. Tooby, eds., 1992. *The Adapted Mind: Evolutionary Psychology and the Generation of Culture*. New York and Oxford: Oxford University Press.

Bateman, A. J. 1948. Intra-sexual selection in Drosophila, *Heredity* 2: 349–368.

Bechtel, W. 2003. Modules, brain parts, and evolutionary psychology. In *Evolutionary Psychology: Alternative Approaches*, eds. S. J. Scher and F. Rauscher. New York: Springer: 211–227.

Beer, G. 2009. *Darwin's Plots: Evolutionary Narrative in Darwin, George Eliot and Nineteenth-Century Fiction*, 3rd ed. Cambridge, UK: Cambridge University Press.

Benbow, C. P. and J. C. Stanley. 1983. Sex differences in mathematical reasoning ability: more facts, *Science* 222: 1029–1031.

The Biology & Gender Study Group. 1988. The importance of feminist critique for contemporary cell biology, *Hypatia* 3: 61–76.

Blecher, S. and R. Erickson. 2007. Genetics of sexual development: a new paradigm, *Am J Med Genet Part A* 143: 3054–3068. Wiley Library Online.

Bleier, R. 1984. *Science and Gender: A Critique of Biology and Its Theories on Women.* New York and Oxford: Pergamon Press.

Bluhm, R., A. J. Jacobson, and H. L. Maibom, eds. 2012. *Neurofeminism: Essays at the Intersection of Feminist Theory and Cognitive Science.* New York: Palgrave Macmillan.

Bronson, F. H. and C. Desjardins. 1976. Steroid hormones and aggressive behavior in mammals. In *The Physiology of Aggression*, ed. K. Moyer. New York: Raven Press.

Buffery, A. and J. Gray. 1972. Sex differences in the development of spatial and linguistic skills. In *Gender Differences, Their Ontogeny and Significance*, eds. C. Ounsted and M. E. Taylor. Edinburgh: Churchill Livingstone.

Buss, D. M. 1999. *Evolutionary Psychology: The New Science of the Mind.* Boston: Allyn and Bacon.

Buss, D. M. and N. Malamuth, eds. 1996. *Sex, Power, Conflict: Evolutionary and Feminist Perspectives.* Oxford, UK: Oxford University Press.

Buss, D. M. and Schmitt, D. P. 1993. Sexual strategies theory: an evolutionary perspective on human mating, *Psychological Review* 100(2): 204–232.

2011. Evolutionary psychology and feminism, *Sex Roles* 64(9): 768–787.

Chargaff, E. 1976. Letter to the editors, *Science* 192: 1448–1449.

Chi, J. G., E. C. Dooling, and F. H. Gilles. 1977. Gyral development of the human brain, *Annals of Neurology* 1: 86–93.

Clarke, J. 1994. The meaning of menstruation in the elimination of abnormal embryos, *Human Reproduction* 9(7): 1204–1207.

Cosmides, L. and J. Tooby, 1992. Cognitive adaptations for social exchange, in *The Adapted Mind: Evolutionary Psychology and the Generation of Culture*, eds. Barkow, J. H., L. Cosmides, and J. Tooby. New York and Oxford: Oxford University Press.

Costanzo, N. S., N. C. Bennett, et al. 2009. Spatial learning and memory in African mole-rats: the role of sociality and sex, *Physiology and Behavior* 96(1): 128–134.

Creager, A. N., E. Lunbeck, and L. Schiebinger, eds. 2001. *Feminism in Twentieth-Century Science, Technology, and Medicine*, Chicago: The University of Chicago Press.

Crouch, N. S., C. L. Minto, L. M. Laio, C. R. J. Woodhouse, and S. M. Creighton. 2004. Genital sensation after feminizing genitoplasty for congenital adrenal hyperplasia: a pilot study, *BJU International* 93(1): 135–138.

Daly, M. and M. Wilson. 2005. The "Cinderella effect" is no fairy tale, *Trends in Cognitive Sciences* 9: 507–508.

Darwin, C. 1859. *On the Origin of Species*. London: John Murray. Reprinted Wildside Press, 2003.

 1860. Letter to Asa Gray.

 1871. *The Descent of Man and Selection in Relation to Sex*. London: John Murray, Albermarle Street.

Dawkins, R. 1978. *The Selfish Gene*. Oxford: Oxford University Press.

 1989. *The Selfish Gene*, 2nd ed. Oxford: Oxford University Press.

Dennett, D. 1995. *Darwin's Dangerous Idea: Evolution and the Meanings of Life*. New York: Simon and Schuster.

Diamond, N. C., G. A. Dowling, and R. E. Johnson. 1981. Morphological cerebral cortical asymmetry in male and female rats, *Experimental Neurology* 71: 261–268.

DeVore, I. and S. L. Washburn. 1963. Baboon ecology and human evolution. In *African Ecology and Human Evolution*, eds. F. C. Howell and F. Bourliere. Chicago: Aldine-Atherton Press.

de Waal, F. 1997. *Bonobo: The Forgotten Ape*. Berkeley and Los Angeles: University of California Press.

Doane, W. W., ed. 1976.(cartoons by B. K. Abbott), *Sexisms Satirized*. Society for Developmental Biology.

Dobzhansky, T. 1973. Nothing in biology makes sense except in the light of evolution, *American Biology Teacher* 35: 125–129.

Douglass, H. 2010. Engagement for progress: Applied philosophy of science in context, *Synthese* 177(3).

 2011. Domestic violence research: Valuing stories. In *Qualitative Criminology: Stories from the Field*, eds. L. Bartels and K. Richards. Leichardt: Australia Hawkins Press, 129–139.

Duhem, P. 1914/1954. *The Aim and Structure of Physical Theory*. Princeton University Press, 2nd ed.

Dukelow, W. R. 1999. Reflections on a century of primatology, *American Journal of Primatology* 49(2): 129–132.

Dunbar, R. I. M. 1988. *Primate Social Systems*. Sydney, Australia: Croom Helm Ltd.

Ehrenreich, B. and D. English. 2005. *For Her Own Good: Two Centuries of the Experts' Advice to Women*, 2nd ed. New York: Anchor Books.

Ehrhardt, A., K. Evers, and J. Money. 1968. Influence of androgen and some aspects of sexually dimorphic behavior in women with the late-treated adrenogenital syndrome, *Johns Hopkins Medical Journal*, 123: 115–122.

Ehrhardt, A. and H. Meyer-Bahlburg. 1981. Effects of prenatal sex hormones on gender related behavior, *Science* 211: 1312–1318.

Eicher, E. and L. Washburn. 1986. Genetic control of primary sex determination in mice, *Annual Review of Genetics* 20: 327–360.

Fausto-Sterling, A. 1985. *Myths of Gender: Biological Theories about Women and Men.* New York: Basic Books.

Fedigan, L. M. 1982. *Primate Paradigm: Sex Roles and Social Bonds.* Eden Press.
Primate Paradigm: Sex Roles and Social Bonds. 2nd ed., with new introduction. University of Chicago Press, 1992.

1986. The changing role of women in models of human evolution, *Annual Review of Anthropology* 15: 25–66.

2001. The paradox of feminist primatology: The goddess discipline? In Creager et al. 2001: 46–72.

Fehr, C. 2001. Pluralism and sex: More than a pragmatic issue, *Proceedings of the Philosophy of Science Association* 68(3):237–250.

2011. Feminist philosophy of biology, The Stanford Encyclopedia of Philosophy (Fall 2011 edition), ed. E. N. Zalta, http://plato.stanford.edu/archives/fall2011/entries/feminist-philosophy-biology/.

Fehr, C and K. S. Plaisance 2010. Socially relevant philosophy of science: An introduction, *Synthese*, 177 (3): 301

Feng, J., I. Spence, and J. Pratt. 2007. Playing an action video game reduces gender differences in spatial cognition, *Psychological Science* 18: 850–855.

Floody, O. R. and D. W. Pfaff. 1974. Steroid hormones and aggressive behavior: approaches to the study of hormone sensitive brain mechanisms for behavior, *Aggression* 52:149–185.

Gastaud, F., C. Bouvattier, L. Duranteau, R. Brauner, E. Thieaud, F. Kutten, and P. Bougneres. 2007. Impaired sexual and reproductive outcomes in women with classical forms of congenital adrenal Hyperplasia, *Journal of Clinical Endocrinology and Metabolism* 92(4): 1391–1396.

Gaulin, S. and D. McBurney. 2000. *Psychology: An Evolutionary Approach.* New York: Prentice Hall.

Geary, M. 1995. An analysis of the Women's Health Movement and its impact on the delivery of health care within the United States, *Journal of Obstetric, Gynecologic, & Neonatal Nursing* Vol. 20 No. 11: 24–31.

Geschwind, N. and P. Behan. 1982. Left-handedness: association with immune disease, migraine, and developmental learning disorder, *Proceedings of National Academy of Sciences* 79: 5097–5100.

1984. Laterality, hormones, and immunity. In *Cerebral Dominance: The Biological Foundations*, eds. N. Geschwind and A. M. Galaburda. Cambridge, MA: Harvard University Press.

Gilbert, S. F. 1988. Cellular politics. In *American Development of Biology*, eds. R. Rainger, K. R. Benson, and J. Maienschein. Philadelphia: University of Pennsylvania Press.

1994. *Postscript to the Importance of Feminist Critique to Cell Biology* (Biology and Gender Study Group 1988).

2014. *Developmental Biology*, 10th ed. Sulnderland, MA: Sinauer Associates.

Gilbert, S. F. and K. A. Rader. 2001. Revisiting women, gender, and feminism in developmental biology. In Creager et al, 2001: 73–97.

Glashow, S. 1987. Quoted in "Does ideology stop at the laboratory door: A debate on science and the real world" in *The New York Times*, Oct. 22, 1989.

Goldberg, S. 1973. *The Inevitability of Patriarchy*. New York: Morrow.

Goodall, J. 1965. Chimpanzees of the gombe stream reserve. In *Primate Behavior Field Studies of Monkeys and Apes*, ed. I. DeVore. New York: Holt, Rinehart & Winston.

Gorski, R., J. H. Gordon, J. E. Shryne, and A. M. Southan. 1978. Evidence for a morphological sex difference within the medial preoptic area of the rat brain, *Brain Research* 148: 333–346.

Gould, S. J. 1980. Sociobiology and the theory of natural selection, *American Association for the Advancement of Science Symposia* 35, 257–69.

Gould, S. and R. Lewontin, 1979. The spandrels of San Marco and the Panglossian paradigm: A critique of the adaptationist programme, *Proceedings of the Royal Society of London, B, Biological Sciences*, 205(1161): 581–598.

Gowaty, P. A., ed. 1997. *Feminism and Evolutionary Theory: Boundaries, Intersections, and Frontiers.*. New York, NY: Chapman & Hall.

2003. Sexual natures: How feminism changed evolutionary biology, Signs 28:3. University of Chicago Press.

Guyton, A. C. 1984. *Physiology of the Human Body*. 6th ed. Philadelphia: Saunders College Publishing.

Haack, S. 1993. Epistemological reflections of an old feminist, *Reason Papers* 18(Fall 1993): 31–43.

Hall, G. S. 1905. *Adolescence Vol. II*. New York: D. Appleton, 588.

Hanson, N. R. 1958. *Patterns of Discovery: An Inquiry into the Conceptual Foundations of Science*. Cambridge, UK: Cambridge University Press.

Haraway, D. 1984. Primatology is politics by other means, PSA: Proceedings of the Biennial Meeting of the Philosophy of Science Association

1989. *Primate Visions: Gender, Race, and Nature in the World of Modern Science*. New York: Routledge.

Harding, S. 1991. *Whose Science? Whose Knowledge? Thinking from Women's Lives*. New York: Cornell University Press.

Hempel, C. 1966. *Philosophy of Natural Science*. New York: Prentice Hall.

Hrdy, S. B. 1977. *The Langurs of Abu: Female and Male Strategies of Reproduction*. Harvard University Press.

1984. Introduction to *Female Primates: Studies by Women Primatologists*, ed. M. F. Small. Alan R. Liss. In Small 1984: 103–109.

1986. Empathy, polyandry, and the myth of the coy female. In *Feminist Approaches to Science*, R. Bleier, ed. New York: Pergamon Press.

1999. *The Woman That Never Evolved*. Cambridge, MA: Harvard University Press.

2009. *Mothers and Others: The Evolutionary Origins of Human Understanding*. Cambridge, MA: Belknap Press of Harvard University Press.

Hubbard, R. 1983. Have only men evolved? in *Discovering Reality: Feminist Perspectives on Epistemology, Metaphysics, Methodology, and Philosophy of Science*, eds. S. Harding and M. Hintikka. Dordrecht: D. Reidel.

Hume, D. 1739. *A Treatise on Human Nature: Being an Attempt to Introduce the Experimental Method of Reasoning into Moral Subjects*, London: Printed for John Noon, at the *White-Hart*, near *Mercer's-Chapel*, in *Cheapside*.

Jolly, A. 1984. The puzzle of female feeding priority. In *Female Primates: Studies by Women Primatologists*, M.F. Small, ed. New York: Alan R. Liss.

Jordan-Young, R. 2010. *Brainstorm: The Flaws in the Science of Sex Differences*. Cambridge, MA: Harvard University Press.

Jordan-Young, R. and R. I. Rumiati. 2012. Hardwired for sexism? Approaches to sex/gender in neuroscience. In *Neurofeminism: Issues at the Intersection of Feminist Theory and Cognitive Science*, eds. R. Bluhm, A. J. Jacobson, and H. L. Maibom. New York: Palgrave Macmillan.

Kansaku, K and S. Kitazawa. 2002. Imaging studies on sex differences in the lateralization of language, *Neuroscience Research* 41: 333–337.

Keller, E. F. 1995. *Refiguring Life: Metaphors of Twentieth Century Biology*. Columbia University Press.

1997. Developmental biology as a feminist cause? *Osiris* 12: 16–28.

Kinsbourne, M. 1980. If sex differences in brain lateralization exist, they have yet to be discovered, *The Behavioral and Brain Sciences* 3(4): 20–36.

Kitcher, P. 1985. *Vaulting Ambition*. Cambridge, MA: MIT Press.

Kuhn, Thomas A. 1962. *The Structure of Scientific Revolutions*. Chicago: University of Chicago Press.

Kunetka, J. 1982. *Oppenheimer: The Years of Risk*. New Jersey, US: Prentice Hall.

Lancaster, J. 1973. In praise of the achieving female monkey. *Psychology Today* 7(4): 20–36.

Laqueur, T. 1987. Orgasm, generation, and the politics of reproductive biology. In *The Making of the Modern Body*, eds. C. Gallagher and T. Laqueur. Berkeley, CA: University of California Press.

Levy-Agresti and J. R. and W. Sperry 1968. Differential perceptual capacities in major and minor hemispheres, *National Academy of Sciences* 61, 1151 (Abstr.).

Lloyd, E. A. 1993. Objectivity and the double-standard for feminist epistemologies, *Synthese* 104(3): 351–381.

Lloyd, E. A. 1996. Science and anti-science: objectivity and its real enemies. In *Feminism, Science, and the Philosophy of Science*, eds. L. H. Nelson and J. Nelson. Dordrecht: Kluwer Academic Publishers.

2003. Violence against science: rape and evolution. In *Evolution, Gender, and Rape*, ed. C. Travis. Cambridge, MA: MIT Press: 235–262.

Longino, H. E. 1990. *Science as Social Knowledge: Values and Objectivity in Scientific Inquiry*. Princeton: Princeton University Press.

1996. Cognitive and non-cognitive values in science: Rethinking the dichotomy. In *Feminism, Science, and the Philosophy of Science*, eds. L.H. Nelson and J. Nelson. Dordrecht: Kluwer Academic Publishers: 39–58.

2002. *The Fate of Knowledge*. Princeton, NJ: Princeton University Press.

Longino, H. E. and R. Doell 1983. Body, bias, and behavior: A comparative analysis of reasoning in two areas of biological science, *Signs* 9(2): 106–127.

Lyell, C. 1830. *Principles of Geology*. London: John Murray, Albemarle Street.

Maccoby, E. and C. Jacklin. 1974. *The Psychology of Sex Differences*. Stanford, CA: Stanford University Press.

Malthus, T. R. 1798. *An Essay on the Principles of Population*. London: J. Johnson.

Marieskind, H. 1975. The women's health movement, *International Journal of Health Services*. 5(2): 217–223.

Martin, E. 1991. The egg and the sperm: how science has constructed a romance based on stereotypical male-female roles, *Signs: Journal of Women in Culture and Society* 16.

2001. *The Woman in the Body: A Cultural Analysis of Reproduction*. Boston, MA: Beacon Press.

Mayr, E. 1972. Sexual selection and natural selection. In *Sexual Selection and the Descent of Man*, ed. B. Campbell. London: Heinemann, 87–104.

1989. *Toward a New Philosophy of Biology: Observations of an Evolutionist*. Cambridge, MA: Harvard University Press.

McGuinnes, C. and K. Pribram 1980. The neuropsychology of attention: Emotional and motivational controls. In *The Brain and Psychology*, ed. M. C. Wittrock. New York: Academic Press.

Merton, R. K. 1968. *Social Theory and Social Structure*, 2nd ed. New York: Free Press.

Milan, E. L. 2012. Making males aggressive and females coy: Gender across the animal-human boundary, *Signs* 37(4): 935–959.

Mill, J. S. 1869. *The Subjection of Women*. London: Longmans, Green, Reader & Dyer.

Nelson, L. H. 1990. *Who Knows: From Quine to a Feminist Empiricism*. Philadelphia: Temple University Press.

1995. A feminist naturalized philosophy of science, *Synthese* 104(3): 399–421.

2003. The descent of evolutionary explanations: Darwinian vestiges in the social sciences. In *The Blackwell Guide to the Philosophy of the Social Sciences*, eds. S. Turner and P. Roth. Oxford, UK: Blackwell: 258–290.

Nelson, L. H. and A. Wylie, eds., 2004. *Hypatia Special Issue: Feminist Science Studies*. Indiana University Press.

Nesse, R. M. and G. C. Williams. 1994. *Why We Get Sick: The New Science of Darwinian Medicine*. New York: Time Books.

Parker, P. and N. T. Burley 1997. *Avian Reproductive Tactics: Male and Female Perspectives, Ornithological Monographs*. Lawrence KS: Allen Press.

Pinker, S. 2009. *How the Mind Works*. New York: W.W. Norton & Co.

Potter, E. 2001. *Gender and Boyle's Law of Gases*. Bloomington and Indianapolis: Indiana University Press.

Profit, M. 1993. Menstruation as a defense against pathogens transported by sperm, *Quarterly Review of Biology* 68(3): 335–386.

Quine, W. V. 1960. *Word and Object*, Cambridge, MA: MIT Press.

1966. Posits and Reality. In *The Ways of Paradox and Other Essays*. New York: Random House: 233–241.

1981. On the nature of moral values. In *Theories and Things*. Cambridge, MA: Harvard University Press.

1987. *Quiddities: An Intermittingly Philosophical Dictionary*. Cambridge, MA: Harvard University Press.

Resnik, D. B. and K. C. Elliott. 2016. The ethical challenges of socially responsible science, *Accountability in Research: Policies & Quality Assurance* 23(1): 31–46.

Richardson, R. 2001. Evolution without history: critical reflections on evolutionary psychology. In *Conceptual Challenges in Evolutionary Psychology: Innovative Research Strategies*, ed. H. R. Halcomb. Dordrecht: Kluwer, 327–373.

Rose, H. 1994. *Love, Power, and Knowledge: Towards a Feminist Transformation of the Sciences*. Bloomington, IN: Indiana University Press.

Rosenberg, C. E. 1979. The therapeutic revolution: Medicine, meaning, and social change in nineteenth-century America. In *The Therapeutic Revolution: Essays in the Social History of American Medicine*, eds. M. J. Vogel and C. E. Rosenberg. Philadelphia: University of Pennsylvania Press, 3–25.

Rosser, S. 1986. The relationship between women's studies and women in science. In *Feminist Approaches to Science*, ed. R. Bleier. Oxford, UK: Pergamon Press.

Rowell, T. E. 1974. The concept of social dominance, *Behavioral Biology* 11(2): 131–154.

Ruse, M. 2008. *Darwinism and its Discontents*. Cambridge, UK: Cambridge University Press.

2012. *The Philosophy of Human Evolution*. Cambridge, UK: Cambridge University Press.

2017. *Darwinism as Religion: What Literature Tells Us about Evolution*. New York: Oxford University Press.

Sade, D. S. 1968. Inhibition of mother-son mating among free-ranging rhesus monkeys, *Science and Psychoanalysis* 12: 18–38.

1972. A longitudinal study of social relations of rhesus monkeys. In *Functional and Evolutionary Biology of Primates*, ed. R. H. Tuttle. Chicago: Aldine-Atherton Press.

Sayers, J. 1982/1990. *Biological Politics: Feminist and Anti-Feminist Perspectives*. New York: Routledge.

Schum, J. E. and K. E. Wynne-Edwards. 2005. Estradiol and progesterone in paternal and non-paternal hamsters (Phodopus) becoming fathers conflict with hypothesized roles, *Hormones and Behavior* 47: 410–418.

Shamoo, A. E. and D. B. Resnick 2014. *Responsible Conduct of Research*, 3rd edition. New York, New York: Oxford University Press.

Schatten, G. and H. Schatten. 1983. The energetic egg, *Sciences* 23(5): 28–34.

Schiebinger, L. 1999. *Has Feminism Changed Science?* Cambridge, MA: Harvard University Press.

2014. *Responsible Conduct of Research*, 3rd edition. New York, New York: Oxford University Press.

Short, R. V. 1972. *Reproduction in Mammals, Book 2*. Cambridge, UK: Cambridge niversity Press.

Synthese 2010. *Special Issue: Making Philosophy of Science Socially Relevant*, eds. K.S. Plaisance and C. Fehr. *Synthese* 177 (3).

Singh, D. 1993. Body shape and women's attractiveness: the critical role of waste-to-hip ratio, *Human Nature* 4:297–3211.

Slocum, S. 1975. Woman the gatherer. In *Toward an Anthropology of Women* ed. R.R. Reiter. New York: Monthly Review Press.

Small, M. F. ed. 1984. *Female Primates: Studies by Women Primatologists*. New York: Alan R. Liss, Inc.

Smith-Rosenberg, C. 1973. Puberty to menopause: the cycle of femininity in nineteenth-century America, *Feminist Studies*, Vol. 1, No. 3/4 58–72.

Smuts, B. 1992. Male aggression against women: an evolutionary perspective, *Human Nature* (3): 1–44.

Sober, E. 1993. *Philosophy of Biology*. Boulder, CO and San Francisco, CA: Westview Press.

Spencer, H. and J. Masters. 1992. Sexual selection: Contemporary debates. In *Keywords in Evolutionary Biology*, eds. E. F. Keller and E. A. Lloyd. Cambridge, MA: Harvard University Press.

Sperling, S. 1991. Baboons with briefcases: feminism, functionalism, and sociobiology in the evolution of primate gender, *Signs* 17:1, Autumn 1991, 1–27.

Stevert, L.L. 2006. *Menopause: A Biocultural Perspective*. New Brunswick, NJ: Rutgers University Press.

   2008. Should women menstruate? An evolutionary perspective on menstrual-suppressing oral contraceptives. In Trevathan et al. 2008: 181–195.

Strier, K. 1994. The myth of the typical primate, *Yearbook of Physical Anthropology* 37: 233–271.

Swedell, L. 2012. Primate sociality and social systems, *Nature Education Knowledge* 3 (10): 84.

Symons, D. 1992. On the use and misuse of Darwinism in the study of human behavior. In *The Adapted Mind: Evolutionary Psychology and the Generation of Culture*, eds. Barkow, J. H., L. Cosmides, and J. Tooby. New York and Oxford: Oxford University Press: 137–159.

Takasaki, H. 2000. Traditions of the Kyoto school of field primatology in Japan. In *Primate Encounters*, ed. S. C. Strum and L. M. Fedigan. University of Chicago Press: 151–164.

Tang-Martinez, Z. 2000. Paradigms and primates: Bateman's principle, passive females, and perspectives from other taxa. In *Primate Encounters: Models of Science, Gender, and Society*, eds. S. C. Strum and L. M. Fedigan. Chicago: University of Chicago Press: 261–274.

Taub, D. 1980. Female choice and mating strategies among wild Barbary macaques. In *The Macaques*, ed. D. Lindburg. New York: Van Nostrand Reinhold: 287–344.

Tavris, C. 1992. *The Mismeasure of Woman*. New York: Touchstone: Simon and Schuster.

The Boston Women's Health Book Collective. 1973. *Our Bodies, Ourselves*. New York: Simon and Schuster.

Thornhill, R. and C. Palmer. 2000. *A Natural History of Rape*. Cambridge, MA: MIT Press.

Trevathan, W. et al, eds. 2008. *Evolutionary Medicine and Health: New Perspectives*. Oxford, UK: Oxford University Press.

Trivers, R. 1972. Parental investment and sexual selection. In *Sexual Selection and the Descent of Man*, ed. B. Campbell. Chicago: Aldine-Atherton: 136–179.

Truth, Sojourner. 1851. Ain't I a woman? Speech delivered at the Women's Convention in Akron Ohio and transcribed by Marius Robinson and published in *The Anti-Slavery Bugle*.

Tuana, N. 2010 Leading with ethics, aiming for policy, *Synthese* 177(3): 471–492.

Tuttle, R. H. 2014. *Apes and Human Evolution*. Cambridge, MA: Harvard University Press.

van Fraassen, B. 1980. *The Scientific Image*. Oxford: Clarendon Press

Wade, N. 1977. *The Ultimate Experiment: Man-Made Evolution*. New York, New York: Walker Publishing Company.

Washburn, S. L. 1961. *The Social Life of Early Man*. New York: Viking Fund Publications in Anthropology.

Washburn, S. L. and C. S. Lancaster. 1968. The evolution of hunting. In *Man the Hunter*, eds. R. B. Lee and I. Devore. Chicago: Aldine-Atherton Press: 293–303.

Williams, G. C. and R. M. Nesse. 1991. The dawn of Darwinian medicine, *The Quarterly Review of Biology* 66(1): 1–22.

Williams, G. C. 1966. *Adaptation and Natural Selection*. Princeton: Princeton University Press.

Wilson, E. O. 1975. *Sociobiology: The New Synthesis*. Cambridge, MA: Harvard University Press.

1978a. *On Human Nature*. Cambridge, MA: Harvard University Press.

1978b. Academic vigilantism and the political significance of Socioibiology, *Bioscience* 26(3): 183–190.

Wrangham, R. W. 1980. An ecological model of female-bonded primate groups, *Behaviour* 75: 262–300.

Wylie, A. 1997. The engendering of archaeology, *Osiris* 12: 80–99.

2003. Why standpoint matters. In *Science and Other Cultures: Issues in Philosophies of Science and Technology*, eds. R. Figueroa and S. Harding. New York: Routledge.

Wyndman et al. 2015. *Social Responsibility: A Preliminary Inquiry into the Perspectives of Scientists, Engineers, and Health Professionals*. Washington, DC: AAAS.

Zihlman, A. 1978. Women in human evolution, Part II, subsistence and social organization among early hominids, *Signs: Journal of Women in Culture and Society* 4: 4–20.

Zuk, M. 1997. Darwinian medicine dawning in a feminist light. In P. Gowaty, ed., *Feminism and Evolutionary Biology: Boundaries, Intersections, and Frontiers*. New York: Chapman & Hall: 417–430.

Index